Cause, Effect, and Control of Accidental Loss with Accident Investigation Kit

· Includes CD-ROM Version Of ·
Accident Investigation Kit

Cause, Effect, and Control of Accidental Loss with Accident Investigation Kit

Ron C. McKinnon, CSP

CRC Press
Taylor & Francis Group
Boca Raton London New York

CRC Press is an imprint of the
Taylor & Francis Group, an **informa** business

CRC Press
Taylor & Francis Group
6000 Broken Sound Parkway NW, Suite 300
Boca Raton, FL 33487-2742

First issued in paperback 2020

ISBN-13: 978-0-367-45550-7 (pbk)
ISBN-13: 978-1-56670-523-3 (hbk)

**Visit the Taylor & Francis Web site at
http://www.taylorandfrancis.com**

**and the CRC Press Web site at
http://www.crcpress.com**

Library of Congress Cataloging-in-Publication Data

Catalog record is available from the Library of Congress.

PREFACE

Accidents have plagued mankind from the beginning of time. At one stage during our evolution, accidents were accepted as being merely part and parcel of life. Many a mother has said to her children after a nasty fall, scrape, or bruise, "Don't worry dear, it was only an accident."

Recently, however, more effort has been centered around the prevention of accidents at work and their accompanying mental and physical anguish and eventual financial loss.

Situation

Even today, safety is measured by the number and severity of injuries. Many people are still convinced that the majority of accidents are caused by unsafe acts. The internationally accepted measure of safety, the disabling or lost-time injury, is unreliable, as it is the end result of numerous luck factors.

Purpose

The purpose of this book is to examine the factors that come together to result in an undesired event (accident) with either potential loss, or actual loss in the form of personal injury, property damage, or business interruption. The book will examine the entire process that constitutes the chain reaction of an accident.

Framework

The framework for the book is the Cause, Effect, and Control of Accidental Loss (CECAL) Sequence, or Domino Sequence, depicting the sequence of events that culminates in loss. This 15-domino loss-causation model was originally proposed by McKinnon in the book *Contemporary Issues in Strategic Management,* edited by Elsabie Smith, Nicholas I. Morgan, et al. The proposed CECAL accident sequence was published on page 283.

The work of some of the world's greatest safety pioneers has been expanded on to compile the CECAL domino sequence. The motivation for the modification, update, and reexamination of the components of an accident was inspired by writings of some of these safety pioneers.

Ted S. Ferry summarized the objective of this book by saying:

> The domino theory is durable. First presented by Heinrich about 1929, it has been updated by several persons. In 1976 Bird used a newer sequence with five dominoes identified as lack of control, basic causes, immediate causes, the incident, and injury to people and/or damage to property. (p.143)

Even today Heinrich's domino sequence is still referred to and used as part of safety-, risk management-, and occupational hygiene-training courses. As with all theories, safety must be kept evergreen to keep pace with changing technology, people and norms in general.

Ferry also mentioned the seven-domino sequence cited by Marcum. He said that in Heinrich's fifth book, Petersen continued the tradition.

Another assumption that has not changed over the years is that the majority of accidents are caused by the unsafe acts of people. Heinrich's loss causation model and his theorems on safety were so radical to the safety industry at the time that readers immediately accepted his statement that 88% of accidents were caused by the unsafe acts of people and this is still believed, incorrectly, today.

Fred A. Manuele PE, CSP, describes this scenario in a nutshell:

> Heinrich's causation model has prominently been used by safety practitioners. Other causation models are extensions of it. However, the wrong advice is given when such models and incident analysis systems focus primarily on: characteristics of the individual; unsafe acts being a prime cause of incidents; and measures devised to correct "man failure," mainly to affect an individual's behavior. 8) Heinrich also wrote, "...a total of 88% of all industrial accidents ... are caused primarily by the unsafe acts of persons." Those who continued to promote the idea that 88 or 90 or 92% of all industrial accidents are caused primarily by the unsafe acts of persons do the world a disservice. (p.31)

Dan Petersen is quoted often in this book and this quotation of his summarizes the purpose of the CECAL study:

> Our present framework of thinking in safety should be examined, and perhaps challenged more than it is. Much of what we do today is based on principles developed long ago.
>
> It may now be time to re-examine those principles and look at some newer principles that have come upon the safety scene. This is the intent of Part One. (p.7)

Importance

The CECAL loss causation analysis of an accident is of vital importance to the safety management profession. It calls for a different way of looking at, measuring, and promoting the prevention of occupational injuries, damage, and disease. This book clearly demonstrates that traditional forms of safety measurement, accident investigation, and the near disregard of near-miss incidents, has to change before the accident toll can be reduced.

Theory

The CECAL sequence proposed that all forms of accidental loss are triggered by the failure to identify the hazards, and to analyze and evaluate the risk and institute risk-control measures. This in turn leads to weaknesses in the management system, which gives rise to basic and personal factors, commonly referred to as the basic causes of accidents. These basic causes prompt unsafe acts to be committed and in turn allow unsafe conditions to be created. Once this situation exists, Luck Factor 1 determines whether there will be a contact with a source of energy. No contact with a source of energy results in a near-miss or, as is defined in this book, an *incident*. Should there be contact with a source of energy, Luck Factor 2 then determines the outcome of the exchange of energy. The outcome could be injury, property damage, or business interruption or a combination of two or all three. If the exchange of energy causes personal injury, Luck Factor 3 then determines the severity of the injury. The 15th domino in the CECAL sequence depicts the costs that are incurred as a result of any undesired event, which results in a loss.

DATA COLLECTION

Data collection for the proposal of the CECAL accident causation sequence has been done by means of:

- research

- case studies

- real life examples

- numerous information sources

Research

The research that has culminated in the CECAL sequence has included 26 years of safety experience by the author as well as some seven years in the electrical contracting industry. During that time in the safety profession, numerous safety surveys, safety gradings, loss control audits, accident investigations, safety training presentations, and consultations were carried out. Thousands of safety interviews were held during this period and the safety systems in existence in more than 850 major organizations were examined, inspected, and audited.

The practical experience gained by the author in countries such as South Africa, Zimbabwe, Zambia, Botswana, Namibia, Canada, Hong Kong, Sweden, Chile, Peru, Australia, United Kingdom, and the United States has culminated in the CECAL theory. Reference is often made to the book *Five Star Safety, an Introduction*, by the author. This is a 500-page manuscript, which is currently being co-published by Ron McKinnon and Dr. Bill Pomfret, director of Safety Projects International, Kanata, Canada. Although not yet published, this book has been used as a valuable source for the CECAL book.

Case Studies

Selective case studies from the author's past experience have been used to emphasize certain aspects of the CECAL theory. Past events, both national and international, have been referred to and the respective sources quoted.

Numerous accidents have been referred to and these include fatal accidents, accidents that have resulted in injury and those that have been termed near-misses. To protect the actual persons involved in these case studies, names, dates, places, and other details have been omitted to avoid any involvement, litigation, or embarrassment. The case studies have been used only to emphasize the relationship between real life situations and the theories proposed by CECAL.

Real Life Examples

As with the case studies, actual life examples have been used to expand on ideas and concepts in the book. The Luck Factors 1, 2, and 3 are certainly radical changes to traditional safety thinking and the author has taken the liberty of using these actual life examples to emphasize the existence of these luck factors.

Actual near-miss reporting systems have been referred to. Once again, the names of the organizations, individuals involved, and other details have been modified for obvious reasons. Some of the examples given have been extracted directly from organizations' safety programs currently in operation as well as from statistics obtained from organizations.

Sources

Numerous sources have been used to research this document and they include books, safety and health publications, and periodicals. More than 50 sources are quoted in the attached reference list. Personal experience has been referred to in a number of instances. Constant referral to the National Occupational Safety Association (NOSA) Five Star Safety System, the International Loss Control Institute, International Rating System, and Safety Projects International Five Star Safety and Health Management System,™ as well as the BHP Copper North America, San Manuel Safety System is also made.

ACKNOWLEDGMENTS

With so much information and work that has culminated in this document, numerous people need to be thanked. Thanks go to Bunny Matthysen, retired managing director of NOSA and inductee into the Safety Hall of Fame, Frank E. Bird Jr., founding director of the International Loss Control Institute, Carel Labuschagne, managing director of International Risk Control Africa, John Bone, managing director of NOSA, the "NOSA boykies" (five star representatives) of BHP Copper North America, San Manuel Mine. Bryan Wollam deserves recognition for the inspiration. Finally, my wife, Maureen (Nakie) McKinnon, who spent numerous weeks typing and editing this manuscript warrants my deep gratitude.

The contents of this document are dedicated to the thousands of men and women who have died as a result of occupational injuries and diseases, and to the millions who have been and are injured every year in industries and mines around the world.

ABOUT THE AUTHOR

Ron C. McKinnon, CSP, is an internationally experienced and acknowledged safety professional. He has been extensively involved in safety research concerning the cause, effect, and control of accidental loss, near-miss reporting, accident investigation, and safety promotion.

Mr. McKinnon received his B.Sc. in Safety Management from Technikon SA, South Africa and a Management Development Diploma (MDP) from the University of South Africa in Pretoria. He received an M.Sc. in Safety and Health Engineering from Columbia Southern University in Alabama.

From 1973 to 1994, Mr. McKinnon was affiliated with the National Occupational Safety Association (NOSA) in various capacities including safety and health training and motivation. He is experienced in auditing safety systems, implementation of safety programs, and production of training films and videos. During his tenure with NOSA he implemented safety systems and training in seven different countries.

From 1995 to 1999, Mr. McKinnon was safety consultant and safety advisor to Magma Copper and BHP Copper, respectively. At BHP Copper he was a catalyst in the safety revolution in the copper industry that resulted in an 82% reduction in injury.

Mr. McKinnon is a professional member of the ASSE (American Society of Safety Engineers) and an honorary member of the Institute of Safety Management. He is currently Safety Consultant and Trainer for ProSET Training, Inc. in Tucson, Arizona.

Ron C. McKinnon, CSP

TABLE OF CONTENTS

Chapter   *Page*

CHAPTER ONE

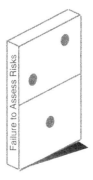

FAILURE TO ASSESS RISKS

Failure to identify the hazards, analyze and evaluate the risk, and set up control measures triggers off a chain of events that lead to accidental loss.

Control

Failure to identify the risks brought about by the business leaves the elements and items that need to be controlled unidentified.

As a result of poor controls, personal and job factors arise that lead to the creation of unsafe work conditions and the unsafe acts of people. These eventually lead to the contact, the property damage, and/or the injury. This chain of events culminates in the last domino in the "Cause, Effect, and Control of Accidental Loss" (CECAL) sequence, which is financial losses. "Effective health and safety management is not 'common sense' but is based on a common understanding of risks and how to control them brought about through good management." (Health and Safety Executive, Great Britain, 1991). Dan Petersen (1998), states that: "If a firm can dictate, in advance, what actions it should take to prevent accidents, then it can measure how well these predetermined actions are executed." (p.37) Risk assessment is a method that is predictive and can indicate potential for loss. With this knowledge an organization is then able to set up the necessary management controls to prevent these risks resulting in losses such as injuries, property damage, business interruptions, and environmental pollution. Many safety programs focus on the consequence of loss and not the control. Effective risk assessment is proactive, predictive safety in the finest form. In risk assessment the keywords are, "It's not what happened, but what **could have** happened."

Vernon L. Grose (1987) talks about the risk management approach as follows:

> The systems approach is a means of looking at even a very large problem in its entirety. You can start by calling whatever you are trying to manage – company, agency, product, family, organization, project, or farm – a system. Then you mark off its boundaries and define its inputs and outputs. (p.11)

Grose says that this allows one to attack all the risks at one time.

Risk

A *risk* can be defined as *any probability or chance of loss*. It is the likelihood of an undesired event occurring at a certain time under certain circumstances. The two major types of risks are speculative risks where there is the possibility of both gain and loss, and pure risk, which offers only the prospect of loss.

Risk Management

Risk management combines the safety management functions of safety planning, organizing, leading, and controlling of the activities of a business so as to minimize the adverse effects of accidental losses produced by the risks arising from the organization.

Physical Risk Management

Physical risk management and financial risk management are the two main components of the science of risk management and the best indicator for safety controls is physical risk management.

Physical risk management consists of identifying the hazards, assessing the risks, evaluating them, and introducing the necessary controls to reduce the probability of these risks manifesting in loss.

Risk Assessment

Risks cannot be properly managed until they have been assessed. The process of risk assessment can be defined as the evaluation and quantification of the likelihood of undesired events and the likelihood of injury and damage that could be caused by the risks. It also involves an estimation of the results of undesired events occurring. One of the biggest benefits of risk assessment is that it will indicate where the greatest gains can be made with the least amount of effort and which activities should be given priority. The safety practitioner now has a prioritization system based on sound risk assessment practices.

Components

Risk assessment has three major components:

1. hazard identification

2. risk analysis

3. risk evaluation

Once the three phases of risk assessment are completed, risk control is then implemented. Risk control can only be instituted once all hazards have been identified and all risks quantified and evaluated. There is a substantial difference between a systematic approach to workplace health and safety and a behavioral-based safety approach. (Jim Howe)(1998, p.20) Howe further says, "The systems approach takes an objective and unbiased view of the workplace by:

1. identifying hazards

2. estimating the level of risk for each hazard

3. controlling hazards according to the hierarchy"

According to Grose (1987), its identification process limits a risk prevention program. "If a risk is not first identified, it can never be evaluated or controlled." (p.13)

1. HAZARD IDENTIFICATION

The first step of a risk assessment is the identification of all possible hazards. A hazard is a situation that has potential for injury or damage to property or the environment. It is a situation or action that has potential for loss.

There are numerous hazard identification methods and techniques. The two main techniques are the comparative and the fundamental methods. Fundamental techniques include Hazard and Operability studies (HAZOP); Failure Mode and Effect Analysis (FMEA); Failure Mode, Effect, and Critical Analysis (FMECA); checklists, and other techniques such as:

- hazard surveys

- hazard indices

- accident reports

- near-miss reports

- critical task identification

- safety audits

Comparative techniques use checklists based on industry standards or existing codes of practice. They could involve comparing the plant in question with similar plants.

HAZOP

According to Grant Purdey, (1996),

> A "HAZOP" involves taking a component (e.g. a valve, or an element of procedure) and stressing it beyond its designing tension and normal operation. In the tradition of HAZOP, nodes of a plant are subject to guidewords applied to relevant physical properties such as flow, temperature and pressure. Then a potential cause of such a deviation is sought, and a consequence defined. If the consequence is undesirable then the hazard has to be addressed by removal, mitigation or control. (p.411)

HAZOP REPORT SHEET

PROJECT: _____

AREA: _____   ITEM/EQUIPMENT:_____

Item No	Guide-word	Possible Causes	Consequences	Action	Person Responsible
1	Position	Moved unintentionally	Bearing no longer aligned	Erect a bump protection device	James Hargrieves

Model 1.1 - A HAZOP Report Sheet showing the guidewords used, possible causes, consequences and mitigating action with responsibility delegated.

FMEA (Failure Mode and Effect Analysis)

The Failure Mode and Effect Analysis is another method of identification of hazards. This method is used for identifying possible failures in the system and resulting consequences. FMEA asks the question, "What system could fail and what would the effect be?"

An example of a FMEA method of hazard identification is given in Model 1.2, which shows extracts of a risk assessment done in a power-generating unit. The FMEA exercise identified the main systems, which could fail within the department and the consequences as a result of main system failures.

What can fail and what will the effect be? (Use normal expected loss).

Item	Effect
The main generator can fail	No electricity generated, no light and heat
The transformer could fail	No power transmission

Model 1.2 - Hazard identification: FMEA.

FMECA (Failure Mode, Effect, and Criticality Analysis)

The Failure Mode, Effect, and Criticality Analysis (FMECA) is a hazard identification method that goes into more depth than the FMEA. The FMECA method examines each component of a system for criticality and identifies the effect on the entire system upon failure of components. This helps focus on critical components within a unit.

Model 1.3 shows an FMECA analysis of a risk assessment. What components are critical to the process and what will the effect be if they fail?

Component	Effect
The brushes could short out	No power would reach the exciter coils
The insulator clip could break	Transformer failure

Model 1.3 - An example of a FMECA analysis.

Event Tree Analysis

The Event Tree Analysis (ETA) is a predictive method of determining the cause and effect of events. The ETA starts with the event and deduces by means of Boolean logic what factors could contribute to the event. Fault Tree Analysis (FTA) is deductive as it deduces the events and sub-events that lead to the main event using the same method as Event Tree Analysis.

Past Accidents and Incidents

A useful method of predicting future hazards is to review past injury and property damage causing accidents as well as near-miss events. By studying past loss-producing events, a pattern can be derived that would indicate certain recurring and inherent hazards within the business.

Near-misses, or events, which under slightly different circumstances could have resulted in a loss, are perhaps the best indicators of the presence of hazards arising from the risks of the business.

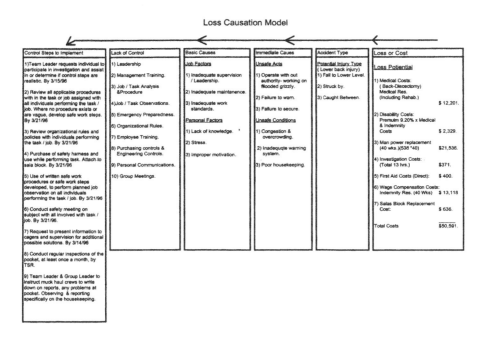

Loss Causation Model

Control Steps to Implement	Lack of Control	Basic Causes	Immediate Causes	Accident Type	Loss or Cost
1)Team Leader requests individual to participate in investigation and assist in or determine if control steps are realistic. By 3/15/96	1) Leadership	**Job Factors**	**Unsafe Acts**	Potential Injury Type (Lower back injury) 1) Fall to Lower Level.	**Loss Potential**
2) Review all applicable procedures with in the task or job assigned with all individuals performing the task / job. Where no procedure exists or are vague, develop safe work steps. By 3/21/96	2) Management Training. 3) Job / Task Analysis &Procedure	1) Inadequate supervision / Leadership. 2) Inadequate maintenance.	1) Operate with out authority- working on flooded grizzly. 2) Failure to warn.	2) Struck by. 3) Caught Between.	1) Medical Costs: (Back-Discectomy) Medical Res. (Including Rehab.) $ 12,201.
3) Review organizational rules and policies with individuals performing the task / job. By 3/21/96	4)Job / Task Observations. 5) Emergency Preparedness. 6) Organizational Rules.	3) Inadequate work standards. **Personal Factors** 1) Lack of knowledge. '	3) Failure to secure. **Unsafe Conditions** 1) Congestion & overcrowding.		2) Disability Costs: Premium 9.20% x Medical & Indemnity Costs $ 2,329.
4) Purchase of safety harness and use while performing task. Attach to sala block. By 3/21/96	7) Employee Training. 8) Purchasing controls & Engineering Controls.	2) Stress. 3) Improper motivation.	2) Inadequate warning system.		3) Man power replacement (40 wks.)(538 *40) $21,536.
5) Use of written safe work procedures or safe work steps developed, to perform planned job observation on all individuals performing the task / job. By 3/21/96	9) Personal Communications. 10) Group Meetings.		3) Poor housekeeping.		4) Investigation Costs: (Total 13 hrs.) $371. 5) First Aid Costs (Direct): $ 400.
6) Conduct safety meeting on subject with all involved with task / job. By 3/21/96					6) Wage Compensation Costs: Indemnity Res. (40 Wks) $ 13,118
7) Request to present information to cagers and supervision for additional possible solutions. By 3/14/96					7) Salas Block Replacement Cost: $ 636.
8) Conduct regular inspections of the pocket, at least once a month, by TSR.					Total Costs $50,591.
9) Team Leader & Group Leader to instruct muck haul crews to write down on reports, any problems at pocket. Observing & reporting specifically on the housekeeping.					

Model 1.4 - Loss causation analysis of a near-miss.

Case Study

The case study on a near-miss incident clearly indicated 10 major hazards that existed that could have resulted in accidental losses in excess of $50,000. These hazards included poor leadership, lack of management training, no emergency preparedness, and little or weak employee training. These hazards extended to the purchasing and engineering controls. This example clearly indicates that near-misses can often highlight hazards that can then be controlled before the loss occurs. Formal and informal incident-recall sessions are imperative if hazard identification is to be done thoroughly. Incident-recall is a method whereby employees recall past near- misses that under slightly different circumstances could have resulted in accidental loss. The losses could cause injury to people, property damage, or interruption to the work process.

Near-misses are also vital in hazard identification. Near-miss reporting systems have often failed in the past yet can be very successful if they are kept anonymous with no repercussion after reporting an incident.

Sample

During the last 20 years I have interviewed more than 500 employees and asked them why near-misses are not reported. The two main reasons that inhibited the reporting were:

1. Fear of disciplinary action should they report an event that might have been caused by an unsafe act on their own or a colleague's part

2. Whenever they report a near-miss they are confronted by, "You saw it, so why didn't you fix it?"

In referring to the accident ratio (Bird and Germain, 1992), and a modified accident ratio it is estimated that nearly 1,000 near-misses could be added to the 600 incidents referred to in the Bird ratio.

One particular factory receives on average some 200 near-misses reported each month. Most of these indicate hazards with potential to cause loss.

Critical Task Identification

A systematic method of listing work tasks and applying a critical task hazard analysis will highlight hazards or hazardous steps within certain tasks. This will enable written safe work procedures to be written for these critical tasks. These will help guide the employee to avoid these risks when carrying out the critical tasks. Past injury and loss experience as well as probabilities are examined. Near-misses are also considered in the critical task evaluation.

Structured safety audits of the existing safety management system also indicate what elements and items are not in control and which pose a hazard.

These audits point out a company's strengths and weaknesses and also indicate which areas may possess more hazards.

Safety Inspections

A thorough safety inspection guided by a hazard control checklist is one of the basic and best methods of identifying physical hazards as well as unsafe practices of people. The hazards, once identified, should be ranked as to their classification by using a simple system such as the A, B, and C rating. An A-class hazard has the potential to cause death, major injury, or extensive business interruption, a B-class hazard has the potential to cause serious non-permanent injury and minor disruption of the business, and a C-class hazard has potential to cause minor injury and non-disruptive business interruption.

2. RISK ANALYSIS

Once hazards have been prioritized in a hazard-ranking exercise, which follows the hazard identification process, the second step in risk assessment is the analysis of the risks.

Risk analysis is the calculation and quantification of the probability, the severity, and frequency of an undesired event as a result of the risk. It is a systematic measurement of the degree of danger in an operation and is the product of the probability and frequency of the undesired event and the resultant severity.

Purpose

The purpose of the risk analysis is to reduce the uncertainty of a potential accident situation and to provide a framework to look at all eventualities. A risk analysis is a risk quantification method that looks not at *what happened* in the past but what *could happen* in the future. It is a method of identifying accidents that have not yet occurred. The risk score is the product of the probability of the event's happening, the severity of the consequences and the frequency of exposure of the event. The probability asks the question, "What are the chances of the event's happening?" The severity asks the question, "If it happens, how bad will it be?" and the frequency asks, "If it happens, how often can it occur and how many people are exposed?" Numbers are allocated to the various degrees of probability, severity, and frequency and the product gives a risk score. In the example (Model 1.5),[5] the probability is ranked on a scale of (1) to (6), the severity on a scale of (1) to (7) and the frequency on a scale of (8) to (1). The highest probability with the worst severity and the most exposure would be 6 multiplied by 7 multiplied by 8, which would be the highest risk score.

Probability

The probability scale used in this risk assessment ranks the probability of injury as a (1) and disaster as (6). The severity is ranked on a scale from (1) for a minor injury, (4) for property damage and (7) for major interruption. The frequency is (8), meaning it could happen 'any time,' to (1) for less often than yearly.

Although there are different scales given to the three dimensions of a risk analysis, a structured approach is taken in quantifying probabilities, frequencies and severity. The example risk analysis form, below, shows the analysis scales.

Analyze the Risk

Probability

What can happen?

Injury	Damage	Fire	Explosion	Catastrophe	Disaster
1	2	3	4	5	6

Severity

How bad could it be?

Minor	Disability	Death	Damage	Extensive Damage	Interruption	Major Interruption
1	2	3	4	5	6	7

Frequency

How often could it happen?

Anytime	Hourly	Daily	Weekly	Monthly	6 Monthly	Yearly	Less Often
8	7	6	5	4	3	2	1

Model 1.5 - An example of the risk analysis.

Risk Matrix

The risk analysis methodology can also be applied to near-miss events by the use of a simplified risk matrix.

A risk matrix is a block diagram with two axes; the loss potential *severity* and *probability* of occurrence both ranked low, medium, medium-high and high. Each undesired event is then ranked as to its potential severity and probability of occurrence. Should the event be in the 'high-high' or 'medium-high – medium-high' area, a thorough investigation of the event is then initiated. Should the ranking be slightly less, a formal inquiry is initiated and should the risk be low, a near-miss incident report is submitted and the problem rectified on the spot.

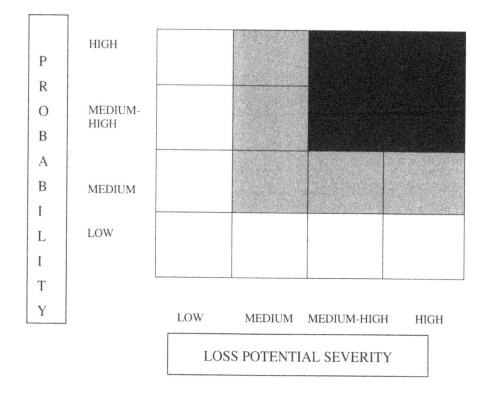

Model 1.6 - A simple risk matrix.

3. RISK EVALUATION

The final step in the risk assessment process is risk evaluation. Grose (1987) explains risk evaluation as:

> Determining the accurate price of a preventive action requires both a skilled estimator and a consistent method of estimating labor costs, capital investments, and expendable outlays – whether one-time, periodic, or continuous. (p.15)

He says that the price tag is critical and is what gets attention. Grose (1987) further explains the purpose of risk evaluation:

> To estimate the value of a preventive action logically, the risk control potential matrix forces consideration of *efficacy* (how much control will result), *feasibility* (how acceptable it will be), and *efficiency* (how much "bang for the buck" will result.)(p.15)

Definition

Risk evaluation could be defined as *the quantification of the risks coupled with an evaluation of the cost of degree of risk reduction and the resultant benefit derived from reducing the risk.*

Objective

The main objective of risk evaluation is to ensure that the cost of risk reduction justifies the degree of risk reduction. Its main aim is to enable management to make decisions on risk reduction priorities.

Risk Free?

For all walks of life, business, and industry to be absolutely risk free would be improbable as well as impracticable. In risk reduction, the objective should be to drive the risk As Low As is Reasonably Practicable (ALARP). The ALARP zone is where the risks have been reduced to where they can be tolerated, as the risk level is acceptable.

If the risks are kept in the ALARP zone it is regarded as normal business practice. Risks that extend above the ALARP zone could prove to be detrimental to the business as well as the employees and others working there.

Model 1.7 shows the risk ranking matrix for six different risks and the risk score on the left-hand side ranging from 0 to 300. This model has been specifically oriented to highlight risks that have the potential for fatalities, and the fatality zone has been determined as being between 100 and 200. The ALARP zone is between 10 and 30 and all risk reduction efforts are made to drive the risks out of the fatality zone and into the ALARP region. This is done with effective safety-and-health management-control systems.

Rank the Risk

Risk

	1	2	3	4	5	6
300						
					201	
200 FATALITY						
180						
160 ZONE		154				
140						
120						119
100	98					
80			77			
60						
40				28		
20 *ALARP						
0						

As Low As is Reasonably Possible

Model 1.7 - Risk Ranking Matrix, showing the fatality zone and the ALARP region.

Cost Benefit Analysis

Risk evaluation could be a type of cost benefit analysis. Purdey (1996) quotes the ICRP (1991), as follows:

> Most decisions about human activities are based on an implicit form of balancing of costs and benefits leading to the conclusion that the conduct of a chosen activity is worthwhile. Less generally, it is also recognized that the conduct of the chosen practice should be adjusted to maximize the benefit to the individual or the society. Cost benefit analysis involves a comparison of the cost of risk reduction measures with the risk-factor cost of the accidents prevented. (p.419)

Some risks have to be tolerated, as reducing all risks totally would prove to be too costly to any business enterprise.

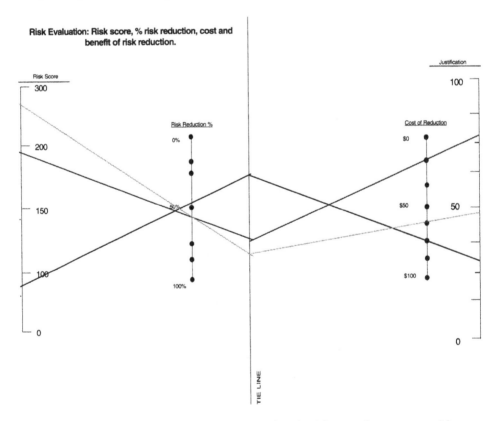

Model 1.8 - A simple Risk Evaluation Model that considers the risk score, the percentage risk reduction, the cost of risk mitigation and plots these variables on a justification scale of 0 to 100. Join the risks score with the percentage reduction and plot a point on the tie line. A 50% risk reduction is recommended initially. Once the cost of a 50% risk reduction is determined, plot this figure on the cost scale. Join the point on the tie line with the point plotted on the cost and draw a line through to intersect the justification scale. Repeat this with the major risks and those that have the highest score on the justification scale should receive priority.

Risk Profile

A risk profile is a model charting the various risks in graph form according to the risk score. The justification scale is imposed on the same model and a second graph, drawn in a different color, shows the justification for rectification as determined by the risk evaluation.

Model 1.9 shows a risk profile for six major risks plotted in solid and the justification plotted in dotted. It is interesting to note that the risks with the highest score in some instances received the lowest justification. Cost of risk mitigation plays an important role in risk evaluation.

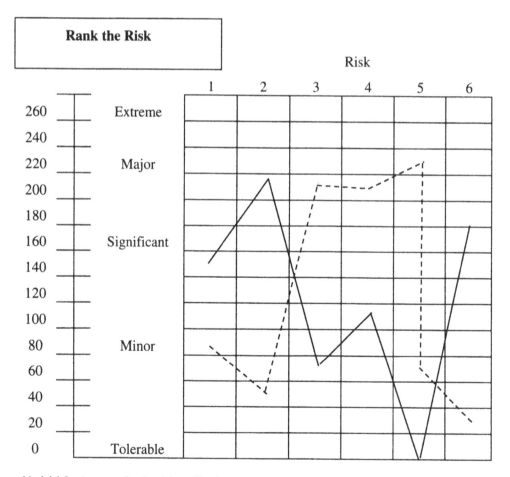

Model 1.9 - An example of a risk profile showing risks score versus justification.

Risk Control

Risks cannot be managed until they have been assessed. Once the risks have been assessed and prioritized through the risk evaluation process, risk controls are now decided upon.

Objective

The objective of risk control is to minimize those risks that have been assessed and wherever possible to transfer them. Exposure to the risk is minimized and contingency plans drawn up to cope with the consequences should the event occur.

Goals

The goals of risk control are to reduce the probability, severity, and frequency of the likelihood of undesired events occurring to a level as low as practical.

Risk Control Methods

There are basically four ways to handle risks:

1. Terminate the risk.

2. Tolerate the risk.

3. Transfer the risk.

4. Treat the risk.

Terminate

The ideal method of risk mitigation is to terminate the risk entirely. This would mean stopping a procedure, changing the business, or disposing of a substance used in the process so that the risk is entirely terminated.

AECI, Modderfontein Dynamite Factory, Modderfontein, South Africa, decided to terminate the risk of manufacturing nitroglycerin-based explosives by ceasing to manufacture that product. This is risk termination.

Tolerate

If risks are in the ALARP region it is acceptable business practice to tolerate these risks, as there is risk in all aspects of business. The tolerating of a risk means that the benefits deriving from the risk outweigh the consequences of the risk.

Transfer

Transferring the risk is not always an ideal solution to control risk and normally involves insuring the risk or placing it in somebody else's hands.

Treat

Treating the risk means setting up safety and health management controls to reduce the risk and therefore reduce the probability of an undesired event occurring.

Treating the risk involves the safety management principle of safety controlling. Safety controlling is defined as the management function of identifying what must be done for safety, inspecting to verify completion of work, evaluating, and following up with safety action.

The acronym IISSMECC is used to explain the control function where:

I = identify and assess the risks.

I = identify the actions needed to reduce the risks.

S = set standards of accountability.

S = set standards of measurement.

M = measure against those standards.

E = evaluate conformances and non-conformances.

C = corrective action to be taken.

C = commendation for work well done.

Summary

As Purdey (1991) states,

> Risk assessment forms a natural part of a good safety management system. It complements and builds upon the existing skills and techniques of the safety professional. In the context of a progressive risk management system, risk assessment is a management tool, which assesses performance, enables analysis, and creates goals and standards. (p.423)

Real Cause

Failure to assess, analyze, evaluate and control risks is the key event that triggers off the chain of events referred to as accidents that result in undesired losses.

Risk assessment, risk management, and risk control is the future of traditional safety as we know it and will lead to predicting where loss-occurring events may happen and enable us to prevent the accidents that have not yet occurred.

In the analysis of a fatal accident (November 1, 1996), the initiating event that caused 10 systems to be out of control was the failure to adequately assess and control the risk.

An investigation of a major equipment damage accident, using an analysis based on the Cause, Effect and Control of Accidental Loss (CECAL) domino sequence, indicated that critical parts, critical tasks, and procedures had not been identified. No risk profile had been compiled for the process and no risk assessment had been carried out. This led to inadequate controls, which then triggered off the sequence of events.

Control

Once the risks have been assessed and the risk control method chosen, safety management *control* is now applied.

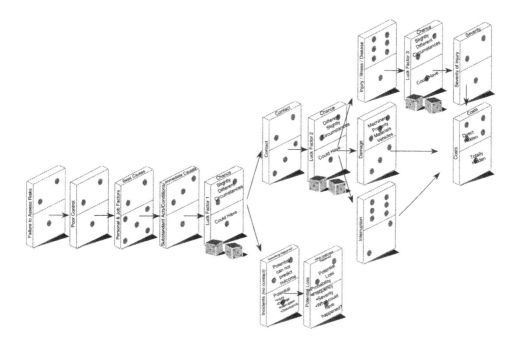

Model 1.10 - The Cause, Effect, and Control of Accidental Loss (CECAL) domino sequence, which shows how the failure to assess the risk triggers poor control and leads to the losses and subsequent costs.

CHAPTER TWO

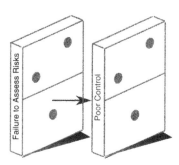

LACK OF SAFETY MANAGEMENT CONTROL

If the risks arising out of the business have not been identified and assessed, they cannot be managed or controlled. This creates lack of, or poor, management control, which is depicted by the second domino in the chain of events leading to undesired events.

> ...there is a great tendency – human tendency – for management to rationalize after experiencing a human tragedy. It is always so much easier to find the "careless acts" on the part of an injured employee which precipitated the accident but, an enlightened management will not hesitate to look beyond the "unsafe act" on the part of an employee and to consider it as a symptom of lack of management control. (Lester A Hudson, 1993)(p.2)

Direction

It has often been said that if you don't know where you are going, any road will lead you there. Safety management is the same. Unless direction is set and actions are based on a thorough understanding of risks, all efforts to reduce accidents may be misdirected.

Alice in Wonderland?

In the classic book, *Alice's Adventures in Wonderland,* (Lewis Carroll), Alice and the Cheshire cat have a conversation:

> "Will you tell me please, which way I ought to go from here?" Alice asked the Cheshire cat. "That depends a good deal on where you want to get to" replied the Cheshire cat. "I don't care much," said Alice after some thought. "Then it doesn't matter which way you go" replied the Cheshire cat. (p.103)

This brief conversation is the same as management not having set controls to combat accidental losses.

Three Control Opportunities

Management has opportunity to control at three stages within the cause and effect sequence.

Pre-contact Control: Pre-contact control is the management work done to prevent accidental loss before it occurs. It consists of all the activities of managing the business and setting up controls in the form of standards and systems to ensure that controls and guidelines are set in place to reduce the probability of the risks manifesting as accidental losses.

Contact Control: Contact control is where the undesired event is not prevented but the consequences of the exchange of energy are minimized. Contact control does not prevent the accident per se but merely minimizes the result. Providing personal protective equipment and ensuring that workers wear it is an ideal example of contact control. Personal protective equipment does not prevent a brick falling from an overhead scaffold but does minimize the amount of energy exchanged between the brick and the employee's head and consequently does exercise a degree of contact control.

Post-contact: Post-contact control includes the management activities that are carried out after there has been an exchange of energy. After a near-miss or loss-producing event, something is done. Examples of post-contact control are:

1. accident investigation

2. revision of procedure

3. injury statistic compilation

4. calculating loss costs

Safety Management

A manager's four main functions, according to Louis A. Allen, are planning, organizing, leading, and controlling. These four main functions can be rewritten to define management's safety-management functions.

Safety Planning

Safety planning is what a manager does to predetermine the consequence of accidents and to determine action to be taken to prevent downgrading events occurring.

Safety Organizing

Safety organizing is the function a manager carries out to arrange safety work to be done most effectively by the right people.

Safety Leading

Safety leading is what a manager does to ensure that people act and work in a safe manner.

Safety Controlling

Safety controlling is the management function of identifying what must be done for safety, setting standards of measurement and accountability, inspecting to verify completion of work, evaluating, and following up with safety action.

IISSMECC

According to *Questions and Answers*, (National Occupational Safety Association)(NOSA)(1994), there are six steps in the controlling function.

1. identification of work situations that involve a high element of risk

2.   setting standards of performance and measurement

3.   setting standards of accountability

4.   measurement against the standards

5.   evaluation

6.   correction (p.13)

Two further steps can be added to this model. They are 'I' for Identification of the risks of the business, which include conducting risk assessments and introducing risk controls to impact on, and direct, the work to be done to minimize risk. The last aspect would be the commendation for compliance to standards and recognition of good performance. The modified model would then look as follows:

I = identify risks via risk assessment.

I = identify the work to be done to mitigate the risks.

S = set standards of measurement.

S = set standards of accountability.

M = measure conformance to standards by inspection.

E = evaluate conformances and achievements.

C = correct deviations from standards.

C = commend for compliance.

Identify Risks

The risk assessment process discussed in Chapter one will ensure that an undertaking has identified all the hazards, analyzed the risks, evaluated them, and ascertained which risk control methods to apply.

Identify Work to be Done

Once the risks have been evaluated and prioritized it is now management's function to identify what work must be done to ensure the treatment of the risks. The risk assessment would have identified both physical and behavioral risks and management can now implement certain control elements to reduce the risk as low as is reasonably practical.

The work to be done to reduce risks is, in many instances, very similar to the basic control activities required by industrial and mining legislation. Examples of the work that may need to be done are the following:

- guarding of all machinery and pinch points

- regular inspections of lifting gear

- safety induction training for new employees

- providing and maintaining of personal protective equipment

- availability of material safety data sheets

- formal accident investigation procedures

- legal injury and disease reporting, etc.

Set Standards of Measurement

It has often been said that managers get what they want and, should management set safety standards, these standards are usually achieved by the organization. Setting standards of measurement clearly indicates how things must be in the work environment. Standards must be in writing and should contain the following headings:

- purpose

- resources

- responsibility

- legal

- general

- monitoring

By setting standards of measurement, management defines the direction in which the organization moves. Should management set a standard for good housekeeping practices, this standard, which is a measurable management performance, is then the way business is done in the future. Standards give the company goals and directions and a definite focus as to the end result. These standards can be of measurement and of performance and ask the questions, "What must the end result be?" and "What must be done by whom?"

An example of a standard of measurement is included in model 2.1. This shows the end result of the standard, which in this case is "scrap and refuse bins: removal system."

Model 2.1 – (NOSA Audit Book, p.25).

Standards can be set for, among other things:

- housekeeping

- demarcation

- hygiene monitoring

- environmental conformance

- safety committees

- plant inspections, etc.

Safety Systems

To guide management in controlling areas of potential loss and to set standards, there are existing safety and health systems that provide excellent frameworks. These are sometimes referred to as "canned" programs. These programs prescribe certain elements under certain headings and give details of what aspects of a safety management program should be instituted as a foundation. Dan Petersen (1998) quotes a 1978 NIOSH study as follows:

> A 1978 NIOSH study identified seven crucial areas needed for safety performance – most of which are not included in packaged programs. The four key elements were: -
>
> 1. top management commitment
> 2. a humanistic approach towards workers
> 3. one-on-one contact
> 4. use of positive reinforcement (p.38)

NOSA Five Star System

The National Occupational Safety Association (NOSA) in South Africa in the early 70s developed the NOSA Five Star System. It was based on 25 years' consulting experience by NOSA field staff and on 150,000 safety surveys that had been conducted across a wide range of industry and mines.

Sections

The NOSA Five Star System consists of five sections:

1. premises and housekeeping

2. mechanical, electrical and personal safeguarding

3. fire protection and prevention

4. incident (accident), recording and investigation

5. health and safety organization

Elements

Falling under these five main sections are 73 safety-program (control) elements, which constitute the basis of any safety and health management system. The elements contain minimum standards for compliance to each element as well as minimum standard detail, which break the element into further achievable objectives.

Audit System

The NOSA Five Star System is a safety audit system. Each minimum standard detail has a point weighting and the entire system totals 2,000 points. Control of these critical elements must have been in operation for at least six months before full cognizance can be given during the auditing process.

Star Grading

To provide encouragement and reward as well as an ongoing challenge, the audit score measured against the achievement of these minimum standards is converted to a percentage and in conjunction with the 12 monthly disabling-injury incidence rate, a star rating is given ranging from 40% for a one-star rating to 61% for a three-star. A rating of 91% or more culminates in a five-star grading, provided the DiiR (disabling-injury rate percentage) is less than one for the 12 preceding months.

Model 2.2 shows an extract from the NOSA Five Star System audit book showing the element, the minimum standard, and the minimum standard detail, as well as the point allocation for those particular standards. To ensure conformity of measurement, the book incorporates an auditor's guide indicating what to look for during the physical inspection, documentation to verify the measurement, and questions to ask to ascertain conformance to the standard.

			QUESTIONS THAT COULD BE ASKED	VERIFICATION	WHAT TO LOOK FOR	NOTES
1.22.4 UNAUTHORISED STACKING	POINTS	ACTUAL			**1.22.4 UNAUTHORISED STACKING**	
Are windowsills clear?	[2]	[ ]		Visual.	Visual check of windowsills and cupboard tops.	
Any stacking on cupboard tops?	[3]	[ ]				
Total	[30]	[ ]				
NS ELEMENT 1.23 FACTORY AND YARD: TIDY						
1.23.1 FACTORY FREE OF SUPERFLUOUS MATERIAL	POINTS	ACTUAL			**1.23.1 FACTORY FREE OF SUPERFLUOUS MATERIAL**	
Are there superfluous materials present?	[5]	[ ]	Are superfluous (re-usable) material lying around? Are redundant (scrap) material lying around?	Visual.	Factory - under roof, scrap, junk, paper, planks, material, etc. lying around. Look behind machines and cupboards.	
Is there any redundant material or equipment?	[4]	[ ]				
Is maintenance and repair work carried out in a neat and tidy manner?	[3]	[ ]				
Are workstations tidy?	[4]	[ ]				
Are racks, shelves and cupboards free of superfluous material and tidy?	[4]	[ ]				
1.23.2 YARD FREE OF SUPERFLUOUS MATERIAL					**1.23.2 YARD FREE OF SUPERFLUOUS MATERIAL**	
Are there superfluous materials present?	[10]	[ ]		Visual.	Yard - all outside areas including the grass areas and roadways.	
Is redundant scrap material properly sorted and stored in designated area?	[5]	[ ]				
1.23.3 CONTROL BY SUPERVISION / HEALTH & SAFETY REPRESENTATIVES					**1.23.3 CONTROL BY SUPERVISION/ HEALTH & SAFETY REPRESEN-TATIVES**	
Is proper control of good housekeeping evident?	[5]	[ ]	Do supervisors/health & safety reps control the areas by reporting? How often do they inspect for good housekeeping?	Look at the inspection sheet of the supervisor/health & safety rep - is housekeeping included? Does the control reflect in the physical conditions?	Controlling areas by reporting (general good housekeeping) Visual check on housekeeping.	
Total	[40]	[ ]				
NS ELEMENT 1.24 SCRAP AND REFUSE BINS REMOVAL SYSTEM						
1.24.1 SCRAP BINS PROVIDED	POINTS	ACTUAL			**1.24.1 SCRAP BINS PROVIDED**	
Are sufficient bins provided?	[5]	[ ]			Number of bins, lids on bins in certain applications, bins adequate.	
Are all bins of the right type/size?	[3]	[ ]		Visual.	Are they the same colour (where possible)?	
Are lids where provided kept in place?	[2]	[ ]				

Model 2.2 – (NOSA Audit Book, p.18).

International Safety Rating System

Another widely used system, which sets standards of measurement, is the International Loss Control Institute's International Safety Rating System (ISRS). This system has also set standards of conformance as well as standards of accountability for safety work. It incorporates physical inspection guidelines and precise instructions to accredited auditors on how to allocate scores. In the ISRS there are 20 main element headings, each one composing a number of elements. The total score is 13,000 points.

Set Standards of Accountability

The next step in the controlling process is the setting of standards of accountability. A standard of accountability indicates *who* will do *what* by *when*. Setting standards of accountability asks, "Who must do it and by when?" An example of setting standards of accountability is the follow-up action after an accident investigation. The control steps to prevent a recurrence are what will be done and this should be followed by a delegation to make one person responsible for the action as well as committing that person to a date for completion.

Control steps to prevent recurrence:	By Whom:	Date to Be Completed
1. John will fix the machine guard	Jonnie Johnson	1/3/99
2. A risk assessment will be done on the Burmaker machine	Al Trios and team	4/4/00
3. Committee to review the standards	Safety Committee	2/3/99
4.		

Model 2.3 - Action steps to prevent a recurrence: a standard of accountability.

Responsibility, Authority and Accountability

There is often confusion about responsibility, authority, and accountability and it is opportune to define these concepts here:

Safety responsibility is the safety function allocated to a post or position.

Safety authority is the total influence, rights, and ability of the post to command and demand safety.

Safety accountability is when a manager is under obligation to ensure that safety responsibility and authority are used to achieve both safety and legal safety standards.

Setting safety standards of accountability involves *who* will do *what* and *when*.

An example of a standard of accountability is the role played by safety and health representatives. Safety and health representatives have been given the authority to inspect their immediate work area using a safety element checklist as a guideline. This inspection is carried out monthly as prescribed by the standard so the standard of accountability is:

1. who? – the appointed health and safety representative

2. Will do what? – will carry out an inspection of his work area

3. When? – this inspection will be carried out on a monthly basis

One of the many safety myths is that "everybody" is responsible for safety. In actual fact, individuals can be *responsible* for only items and people over whom they have *authority* and can thus be held *accountable* for only those conditions and people over whom they have authority.

Measurement

This control function is when management measures what is actually happening in the workplace against the preset standards. To measure successfully, a walkabout inspection must be carried out and people doing these inspections should be aware of, and familiar with, the standards. One of the greatest failings in most safety systems is insufficient or inadequate inspections.

Systems to enable ongoing measurement against standards are part of a safety system and these could include the monthly inspections of local work areas by appointed safety and health representatives. Safety competitions and housekeeping competitions also allow opportunity for measurement against standards.

The setting of standards and constant measuring against those standards immediately identifies strengths and weaknesses of the safety program in place. Safety personnel should also conduct formal inspections on a regular basis and compare with the standards. A checklist should always be used when doing these inspections as it will serve as a constant reminder of what must be measured.

It should be emphasized here that this form of safety management control and the measurement phase of the control process are not merely measuring and comparing injury statistics with other companies or industries. This is a pure management function of measuring whether the organization is living up to the norms agreed to by management and employees in the form of safety standards.

Critical Task Observation

Another form of measurement against standards is critical task observation. This involves observing an employee carrying out a critical task while following the steps of the written safe work procedure. The written safe work procedure is the end result of a critical task identification (task risk assessment) and the critical task analysis process. The observation allows for measurement of workers' performance during the critical task against the prescribed performance dictated by the procedure. The procedure is a standard of performance.

Safety Evaluation

The evaluation process is the quantification of degree of conformance to the standards set. Evaluation of achievement of standards is normally facilitated through the audit process. As Dan Petersen (1998), stated:

> Consequently, the self-build audit – one that accurately measures performance of a firm's own safety system – was viewed as the answer. To construct such an audit, a firm must define:
>
> 1.   Safety system elements;
> 2.   The relative importance of each (weighting);
> 3.   Questions to determine what is happening. (p.38)

Both the NOSA Safety System and the International Safety Rating System are audit-based systems, which systematically quantify the degree of compliance to standards. They evaluate the management work being done to combat loss. Peter Drucker once stated that what gets measured gets done and consequently the evaluation of compliance to safety standards gives an indication of what is being done and what is not being done. The quantification of safety controls is far more reliable and significant than the measurement of safety consequences, which are largely fortuitous.

External Audits

A number of safety, health, and risk-management organizations provide auditing services for clients. This external audit is of tremendous value to any organization as it is totally impartial and conducted by auditors who are thoroughly familiar with the audit protocol. The audit of an organization's safety system is a structured approach to the quantification of safety compliance and adheres to the following sequence:

1. pre-audit meeting

2. audit facilities

3. audit team

4. physical inspection

5. compliance audit

6. systems audit

7. documentation review

8. verification of disabling injury incidence rate

9. management close-out and audit results

Retrospective

The audit results and percentage achievement is normally based on the safety program's achievements during the preceding 12 months. Credit is not given for good intentions but rather for action and activities that have been in operation for at least six months.

Internal Audit

Internal audits of the entire safety system to evaluate conformance with standards should be carried out every six months. The internal audit system should follow the same guidelines as an external audit and will culminate in a percentage conformance as well as a breakdown of conformance against standards for each element.

Section 2.00

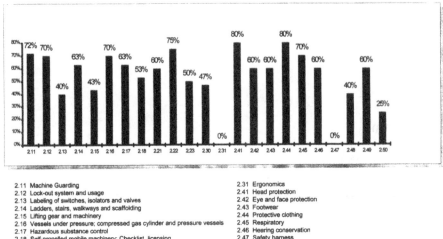

2.11 Machine Guarding
2.12 Lock-out system and usage
2.13 Labeling of switches, isolators and valves
2.14 Ladders, stairs, walkways and scaffolding
2.15 Lifting gear and machinery
2.16 Vessels under pressure; compressed gas cylinder and pressure vessels
2.17 Hazardous substance control
2.18 Self-propelled mobile machinery: Checklist, licensing
2.21 Portable electrical equipment
2.22 Earth leakage relays: use and check
2.23 General electrical installations and electrical machinery
        in hazardous locations
2.30 Handtools: e.g. hammers and chisels

2.31 Ergonomics
2.41 Head protection
2.42 Eye and face protection
2.43 Footwear
2.44 Protective clothing
2.45 Respiratory
2.46 Hearing conservation
2.47 Safety harness
2.48 Hand protection
2.49 Control over personal protective equipment
2.50 Notices and signs: electrical, mechanical, protective equipment,
        traffic signs, symbolic safety signs

Model 2.4 - Shows an example of part of an internal audit report with the percentage conformance, which was evaluated during the audit. This immediately gives management an indication as to the strengths and weaknesses of that safety and health management system. It also indicates where conformance and non-conformances to standards lie.

Weighting Systems

Although traditional weighting systems allocate different numbers of points for each minimum standard detail, the weighting may not be relevant in certain industries and, under certain circumstances, could weaken the intent of the audit process. To weigh each element – minimum standard and minimum standard detail – equally, would be an ideal situation. Some safety systems have incorporated a 0 to 5 rating for each minimum standard detail. This rating implies that each minimum standard detail is as important as the next. The scale indicates 0% compliance to 100% compliance.

Corrective Action

Corrective action is the safety management work that must be done to correct those activities that were not completely controlled. If any critical element is evaluated at less than 100%, some action needs to be taken to ensure total conformance with standards. Management must do what it says it is going to do. The safety standards indicate what must be done. Deviations indicate that the safety objective has not yet been achieved.

Accident Investigation

An accident is caused by a failure in the management system and, after an investigation, certain action plans or controls are recommended to prevent a recurrence of a similar accident. This is corrective action and should be directed at the root cause of the problem, not merely the symptoms. Correcting unsafe acts and unsafe conditions may provide temporary relief but the real cause must be identified and the problem solved.

Common Mistakes

The two most commonly used solutions after an accident are (1) to train people and (2) to write a procedure. These actions may sound very effective but they do not necessarily provide a complete cure for the problem. The reasons for the deviations from standards should be first identified and then corrective action instigated to fix the problem.

Action plans should be drawn up and they should be specific and contain the following directives:

- What must be done?

- Who must do it?

- By when or how often must it be done?

- When must a reevaluation be conducted to ensure conformance?

Commendation

One of the main failings in numerous safety systems and programs is the lack of commendation and recognition. Commendation should be given on achievement of objectives. If a department meets and maintains the housekeeping standard, for example, the entire group should be commended. Commendation for pre-contact safety activities is far more effective than commendation for injury-free periods.

Punishment

The accident sequence indicates a breakdown in the management control system. Weak management controls are normally as a result of failing to assess and manage the risks. Control is a management function and is the most important function to reduce the probability of undesired events, such as accidents, occurring.

As Frank Bird (1992) states, "Punishment as a last resort, but it must be done in a way that communicates your genuine concern." (p.52)

All too often supervision resorts to traditional management styles and institutes disciplinary actions in the form of verbal or written warnings or more drastic punitive measures. As is clearly stated in the management principle of *definition,* solutions to problems can only be prescribed once the real cause of the problem has been identified.

Examples

Good examples of safety control elements are written safe work procedures and planned job observations. Model 2.5 shows a safe work procedure for a hazardous task. The steps appear in a logical sequence, the critical points, which are items to be emphasized, and which have been derived by the critical task analysis, are reproduced in the center column. The observation notes are written opposite each step during the observation. This control procedure not only checks the performance of the person but also the adequacy of the procedure.

WRITTEN SAFE WORK PROCEDURE/PLANNED JOB OBSERVATION				

BUSINESS TEAM:
DEPARTMENT:

HAZARDOUS TASK DESCRIPTION:

WRITTEN SAFE WORK PROCEDURES			PLANNED TASK OBSERVATION	
NO.	STEPS (in sequence)	CRITICAL POINTS (item to be emphasized)	COMMENTS/REASONS OR DEVIATION FROM STANDARDS	NO.
6.	TWIST THE ENDS OF THE EXPOSED WIRE 5-6 TIMES IN A CLOCKWISE MOTION. REPEAT THIS STEP FOR ALL COLORED WIRES	USE CAUTION WHEN TWISTING		
7.	REMOVE ONE KNOCK-OUT FROM PVC SWITCHBOX. USE SCREWDRIVER.	USE CAUTION WHEN REMOVING KNOCKOUT		
8.	INSERT ONE END OF THE ELECTRICAL WIRE THROUGH KNOCK OUT HOLE OF PVC SWITCH BOX	NONE		
9.	CONNECT THREE WIRES FROM ELECTRICAL CABLE LOCATED INSIDE OF THE SWITCHBOX TO THE DUPLEX RECEPTACLE IN ACCORDANCE WITH WIRING DIAGRAM (ATTACHMENT A)	MUST FOLLOW WIRING DIAGRAM		
10.	WIRES MUST BE INSERTED UNDER THE SCREW WITH THE END OF THE WIRE FACING IN A CLOCKWISE DIRECTION.	MUST FOLLOW WIRING DIAGRAM		
11.	USING SLOTTED SCREWDRIVER, TIGHTEN SCREWS TO HOLD COLORED WIRES IN PLACE	NONE		

Model 2.5 - Shows a safe work procedure for a hazardous task.

Model 2.6 is another control system, the safety representative's monthly inspection. Using a checklist of certain critical elements that need to be controlled, the appointed and trained representative inspects his immediate work area and records any deviation from standards. If there are deviations from standards, this report generates a work order to correct the deficiencies.

SAFETY SYSTEM
SAFETY REPRESENTATIVES'
MONTHLY INSPECTION REPORT

LOCATION:_____ NAME:_____

DATE:_____

STANDARD:	YES	NO	COMMENTS:	MAJOR THREAT
1.11 Structure, furniture, and equipment in good and safe condition				
1.11 Floors- clean and free of dirt or other dangers.				
1.12 Good lighting, natural or artificial				
1.13 Ventilation adequate				

Model 2.6 - An example of a safety representative's checklist.

Another pre-contact control in the form of a standard is the number of job observations that are carried out each month. Bearing in mind that these observations are checking both the procedures and performances of critical tasks, this activity is vital, as it has already been determined that these tasks are the critical few i.e., those with greatest potential for loss.

Fatal Accident

In investigating a fatal accident the following appeared in the report:

> Control measures in the form of standards, procedures and monitoring of compliance to procedures are aimed at eliminating the basic cause of an accident in an effort to eliminate the unsafe acts and unsafe conditions. Instituting positive control measures will ensure that the same type of undesired event is eliminated. Positive action is required and ongoing monitoring of these actions is essential. (p.12)

The results of the same accident investigation indicated that there were some 16 elements, minimum standards, and minimum standard details, which were not controlled. The report concluded that if these control elements had been correctly implemented, the accident would most likely not have occurred.

Commendation

If, on the other hand, as a result of the measurement and evaluation, a high degree of conformance to performance standards is found, commendation should be given. Safety, as a profession, has often been guilty of emphasizing the lack of control and not complimenting where good control exists.

Recognition

People at workplaces and in other walks of life thrive on recognition and acknowledgment. Recognizing and acknowledging people for safety work done to prevent loss should be done as often as possible. Traditionally, safety recognition is only given to individuals for being "injury free" or for having worked a certain number of days without lost time injury and, as will be explained, that could just be the result of good luck.

Maintaining good housekeeping, carrying out monthly inspections of ladders, portable electrical equipment, hand tools, personal protective equipment, etc., is an ongoing control system and this effort should be recognized.

Safety and Health Representatives

Safety and health representatives have an extremely difficult task in identifying and reporting deviations from standard and in having to compile a monthly report. Acknowledging their input recognizes this ongoing effort and meeting of the standard.

Internal housekeeping competitions, which acknowledge the winning team for its ongoing efforts, are also great motivators and are part of the control function.

Five Star Rating

The NOSA Five Star Rating is an example of commendation of conformance to established standards.

Management

It is good management practice to commend people for their safety effort. Bearing in mind that control is pre-contact control, this is more important than the recognition of no adverse consequence in the form of severe injury.

Summary

According to Frank E. Bird (1992), lack of control can be grouped into four main areas:

1.   An inadequate safety program

2.   Inadequate program standards

3.   Inadequate compliance with standards

4.   An inadequate management system (p.30)

Safety control eliminates the basic causes of accidents by setting up management systems, and by delegating safety responsibility. This system creates a work environment in which personal factors and job factors are reduced, consequently reducing the unsafe act and the unsafe condition. As Dr. Mark A. Friend (1997) says, "Only members of the management team can create or change the environment. (And it is, after all, their job to do so)." (p.34)

Basic or Root Causes

Inadequate or nonexistent controls of a safety management system constitute the root cause of all accidents.

CHAPTER THREE

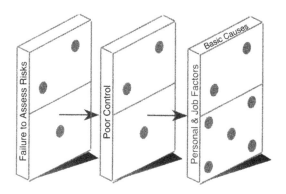

BASIC CAUSES OF ACCIDENTAL LOSS

The lack of, or inadequate, management control, caused by a failure to identify, assess, and control risks, creates basic causes of undesired events. These basic causes are a vital link in the chain reaction that results in an undesired event with loss.

Definition

The management principle of definitions states that a logical and proper decision can only be made when the basic or real problem is first identified. In safety, very seldom is the real problem in the form of basic causes identified and treated. Most safety systems are reactive, as they treat the symptoms and not the causes of accidental loss. It is easier to treat the symptom because the symptom is more readily identifiable, whereas in-depth delving and investigation, coupled with tenacity and risk taking, are needed to unearth the basic causes.

As Dan Petersen (1997) wrote, "By focusing on unsafe acts and conditions, safety managers end up dealing with accidents on a symptomatic level rather than a causal level." (p.39)

Example

After presenting an eight-hour safety-management-training course, one of the attendees, a senior supervisor with many years of experience, approached me with a concerned look on his face. He started to tell me that the employees in his workshop had a bad attitude toward safety. Rather startled by his statement, I then inquired what he meant by a poor attitude to safety. He said that his employees constantly spat on the floor of the workshop with total disregard. I was concerned by his problem and offered to visit the workshop with him. Upon entering the workshop I was struck by the total disorder that confronted me. The lighting was poor, the ventilation systems malfunctioned, and the entire floor was almost ankle deep in debris, dirt, superfluous material, filth, and grime. There was no demarcation of work areas or stacking areas and the floor was almost like a minefield with obstructions. Trip and slip hazards were everywhere. I turned to the supervisor and told him that I felt he was fortunate that his employees were only "spitting" on the floor. It was obvious that the basic cause of their actions was the work environment, which encouraged them to have no respect for their surroundings. In this case, the spitting on the floor was only a symptom of the basic cause of the unsafe and unhygienic work conditions in which they worked.

Heinrich et al. (1959) acknowledges the basic causes of accidents as follows, "While this is an important function, we now know that long-range improvements can best be made by identifying and correcting the basic causes. These can be loosely grouped into 3 categories:

- Management Safety Policy and Decisions

- Personal Factors

- Environmental Factors" (p.35)

Unsafe Behavior

Dr. Mark A. Friend (1997) wrote, "Employees do what seems rational. When they take short cuts or refuse to wear personal protective equipment, it is because management has created an environment in which such actions are the rational way to respond." (p.34)

This statement is perhaps one of the most important, controversial, yet true statements in the science of safety management. Over the years, both safety practitioners and management have been convinced that accidents are largely due to the unsafe actions of people. By focusing on the behaviors of people they have tended to lose sight of the fact that the behaviors are merely a result of the climate that management creates, both within the work environment and by the work-safety ethics that it dictates (or should dictate).

Disease

Bird and Germain (1992), refer to the basic causes as the disease or real cause behind the symptoms. They motivate this by saying:

> This is because the immediate causes (the symptoms, the sub-standard acts and conditions) are usually quite apparent, but it takes a bit of probing to get at basic causes and to gain control of them. Basic causes help explain why people perform sub-standard practices. (p.27)

Example

An example of basic cause analysis is a quotation from a property damage accident (1992), which, under the heading of basic cause analysis, reads as follows:

> Basic or root causes are the real causes of the problem and give rise to the unsafe acts and conditions which are the symptoms. Preventative measures must address the basic causes so that the symptoms do not re-occur. Each unsafe act and condition is repeated here and the basic cause analysis done by asking the questions, why?, why?, why? (p.3)

Three Main Basic Causes

Most safety professionals classify basic causes into personal factors, job factors and natural factors.

Examples of personal factors are:

- inadequate physical or physiological capability

- inadequate mental/psychological capability

- physical or psychological stress

- mental or psychological stress

- lack of knowledge

- lack of skill

- improper motivation

The eight job factor classifications are:

- inadequate leadership and/or supervision

- inadequate engineering

- inadequate purchasing

- inadequate maintenance

- inadequate tools and equipment

- inadequate work standards

- wear and tear

- abuse or misuse

Basic Cause Analysis Example

In investigating a major property damage accident (1997) it was discovered that the unsafe act of "failure to warn" had been committed. Further investigation revealed the reason for this "failure to warn" was "the wrong reading was given by the computer." This in turn did not warn the hoist man as to the true position of the skips (elevators). There was no other verification or check and balance to warn the hoist man that there was imminent danger.

Another unsafe act was that the operator operated the hoist at full speed (operating at unsafe speed). Following a basic cause analysis, the reason for this was found to be "an inadequate or non-existent procedure, which allowed the operator to operate at full speed before confirming the skip's position." (p.4)

Accident Prevention

NOSA (1988) asks the question,

> To prevent the same accident being repeated would it be sufficient to eliminate the immediate causes? Example, the unsafe act and/or the unsafe condition?

> Answer: No! The unsafe acts and/or unsafe conditions (immediate causes) exist because of less obvious causes, (personal and job factors), which must be identified and eliminated. (p.20)

> An example is given as follows:

> Oil on the floor can cause accidents and injury. Identifying the oil as the immediate cause of the accident is correct but if we wipe up the oil we are not treating the basic cause of the oil being on the floor. If we investigate further we may find a fork truck leaking oil, which is the basic cause. To eliminate this (basic cause) the fork truck should be repaired. (p.20)

Example 1

During the investigation of an accident that resulted in injury, the technique of basic cause analysis was applied. Seven unsafe conditions were identified as well as five unsafe acts. In sitting with the team and by using brainstorming techniques, a basic cause analysis indicated that there were some 13 basic causes that had led to the unsafe conditions and unsafe acts.

Example 2

Basic causes create unsafe acts and unsafe conditions. These are the immediate causes of the contact that causes the loss. Model 3.1 shows an accident investigation analysis using basic cause analysis. The basic causes were six job factors and seven personal factors. These created the conditions and actions that caused this loss-producing event.

BASIC CAUSE ANALYSIS	IMMEDIATE CAUSES
Personal Factors:	**Unsafe Acts:**
• Driver was exposed to undue demands resulting in stress. (Supervisor's expectations)	• Driver operating equipment in poor state.
• Improper motivation to save time and effort.	• # 5 Tow Motor was not tagged in poor condition, and in use.
• Lack of skill:	• Joseph (Keith's partner) demand to operate equipment on bad order.
• Inadequate initial instruction.	
• Inadequate practice. (#5 Motor)	• Mechanics inverting brake adjustment cross arm assembly, resulting in assembly bolts facing up.
• Infrequent performance of task. (Tow Motor)	
• Lack of coaching. (Tow Motor)	• #5 Tow Motor equipment check card not properly filled out.
Job Factors:	
• Management does not plan for replacement equipment, in the event of mechanical breakdown, resulting in Electricians' desire to keep damaged equipment to get job(s) done.	**Unsafe Conditions:**
	• # 5 Battery Motor left in unsafe condition not locked out.
• Inadequate employee performance measurement and evaluation. (Proper use of barrier)	• Poor communication between "C" shift and "A" shift workers.
	• Decline barrier was left in improper position.
• Inadequate assessment of maintenance needs and engineering evaluation of changes. (Normal practice to invert cross arm brake adjustment bar)	• Battery of # 5 Tow Motor was setting on the brake assembly bolts, resulting in inoperable

• Inconsistent use of standards, policies and rules combined with inadequate monitoring of standards compliance. (Reporting damages – electricians and dispatcher)	brakes. • #5 Tow Motor missed scheduled maintenance. (Mechanic on maternity leave)
• Inadequate work planning or programming. (Vacation, replacement personnel)	• Tow truck qualification does not always include "Tow Motor."
• Inadequate assessment of loss exposure – engineering. (2% slope)	• 2% slope.

Model 3.1 - The immediate causes and the basic causes of an undesired event.

Magic Words

In endeavoring to uncover the basic causes of any loss causation situation, the immediate causes should first be identified via a thorough investigatory process. Only once the immediate causes are identified can basic causes be derived.

The best and simplest method of deriving basic causes is to ask why? why? why? This basic cause analysis process may result in identifying other unsafe acts or unsafe conditions. These should be further analyzed in an effort to derive the basic causes of why the condition existed or why the behavior was carried out.

Fatal Accident Investigation

During the investigation of a fatal accident, one of the unsafe acts committed was "failure to make secure." By applying basic cause analysis, four personal factors were derived. They were (1) abuse or misuse, (2) stress, (3) lack of experience, and (4) lack of coaching. A further analysis was then done to uncover the reasons for the root causes. In this particular accident investigation the reasons were:

Abuse or Misuse – there were two apparent reasons for this.

1. Convenience, which allowed the job to proceed even though the poles were not compacted.

2. Habit. Site inspections revealed that pole no. 4 was not compacted and several others had also not been correctly compacted.

Stress – five of the witnesses stated that they were under pressure to complete the work.

The example shows that there is always a reason for an unsafe condition or an unsafe act. These reasons must be identified before the undesired event, rather than after. Conducting risk assessments and setting up controls to eliminate basic causes is the best method of loss prevention. An important post-contact control method is accident investigation, which has the main objective of deriving and eliminating the basic cause of the accident to prevent future recurrence. This is retrospective and reactive rather than proactive.

Example 3

During a serious accident investigation (1996) the unsafe act of "ignoring rules" was identified as one of the immediate causes of the accident. This occurs frequently in industry and is often identified as the one and only reason that the event occurs. In this particular instance, the personal factors were sought, and no fewer than eight reasons that the rules were ignored were uncovered. They included:

1. The rules had only been issued 2 weeks prior to the accident.

2. The operator in question was not in possession of the rules.

3. The rules had been blatantly and often ignored in the past with no consequence.

4. There were no inspections during the operation.

5. There was no control of contractors.

6. There was no follow up to ensure compliance with rules.

7. The rules were found to be conflicting.

8. The instructor spent 90% of his time operating and only 10% in training (derived by personal interview).(p.12)

During the same investigation it was discovered that the operator had operated at an unsafe speed. The analysis indicated seven basic reasons that this operator was in a hurry. They included:

1. It was nearing the end of the 12-hour shift.

2. The light was fading quickly (personal observation showed that it would have been dark within the next 15 minutes, making the task impossible).

3. The particular job had been delayed for some 8 hours.

4. Improper motivation.

5. Extreme judgment, decision demands.

6. Confusing demands.

7. Inadequate instruction (these personal factors will not be explained in more detail at this stage, but were what emerged as a result of the analysis).

Can of Worms

In attempting to derive basic causes, the investigator may open a can of worms because traditionally, people were blamed for committing unsafe acts. Certain organizations have made very generalized statements, which have unfortunately been believed and practiced by management, incorrectly, over a number of years. Du Pont states in its training course that 96% of accidents are caused by the unsafe acts of people. On questioning this statement, the company says that people create unsafe conditions anyway, so that means people are guilty of creating unsafe conditions and of committing unsafe acts. Unfortunately, this statement and argument is far too simplistic and far too generalized to enable us to clearly identify and analyze what causes accidents. This attitude also blinds us to the fact that there are reasons that people do what they do and that the work conditions are below organizations' accepted standards concerning housekeeping, neatness, and safety.

Investigating why unsafe acts are normally committed results in questioning the supervision and management control that existed (or did not exist) at the time of the event. This is what opens the first can of worms.

Four-Finger Rule

The four-finger rule is perhaps the best method of explaining how basic causes form an important part of the loss causation process. Immediately upon occurrence of an accident the tendency is to point a finger at the injured person and say, "He messed up," or, "Employee failed to do ..." While pointing a finger, it is interesting to note the position of the other three fingers. They are normally pointing directly back to management, asking "Were we not responsible for this man's action? Have we set the standards, have we monitored the standards, is our behavior beyond reproach?" Answers to these questions may very well reflect that the basic causes are poor or inadequate management control. This leads to another can of worms being opened. Most safety professionals are hesitant to embark on this risky endeavor. Management has been lulled into complacency by such myths as (1) the injured person is the one who messed up, and (2) everybody is responsible for his or her own safety and the safety of others. Thus, management is totally innocent, or so it thinks. It has often been said that if you put a man into a pigsty he is going to behave like a pig, and yet, when he does, we chastise him without asking why the work area is a pigsty in the first place.

It is management's prerogative to set up supervisory controls, systems, and standards and to take the lead to ensure that the basic causes do not exist.

Fault Finding

Failing to identify and rectify basic causes before they result in a loss is relying on luck to prevent the next accident. Basic causes must be eliminated by adequate controls. People are human beings and have made mistakes in the past and will continue to make mistakes because of the very nature of humanity. Setting the controls will help create an environment in which people are less inclined to make mistakes (mess up).

LACK OF CONTROL	BASIC CAUSE ANALYSIS

• Non compliance to standard reporting and correction of equipment in bad order.
 	
• Inadequate training program standards that generalize training for different equipment of the same type.
 	
• Non-compliance to standard reporting damage and operation of barrier.
 	
• No risk-assessment standards.
	Personal Factors:
	
	• Driver was exposed to undue demands resulting in stress. (Supervisor's expectations)
	
	• Improper motivation to save time and effort.
	
	• Lack of skill:
	
	  • Inadequate initial instruction.
	
	  • Inadequate practice. (#5 Motor)
	
	  • Infrequent performance of task. (Tow Motor)
	
	  • Lack of coaching. (Tow Motor)
	
	Job Factors:
	
	• Management does not plan for replacement equipment in the event of mechanical breakdown, resulting in electricians' desire to keep damaged equipment to get job(s) done.
	
	• Inadequate employee performance measurement and evaluation. (Proper use of barrier)
	
	• Inadequate assessment of

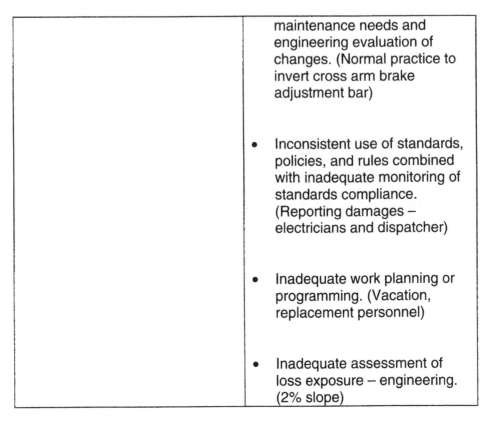

	maintenance needs and engineering evaluation of changes. (Normal practice to invert cross arm brake adjustment bar)
	• Inconsistent use of standards, policies, and rules combined with inadequate monitoring of standards compliance. (Reporting damages – electricians and dispatcher)
	• Inadequate work planning or programming. (Vacation, replacement personnel)
	• Inadequate assessment of loss exposure – engineering. (2% slope)

Model 3.2 - Basic cause analysis.

Example 4

Model 3.2 is an accident loss causation analysis once again showing basic causes of a breakdown in the system that led to loss. In this example, some of the basic causes were identified as poor work planning, lack of motivation, stress, inadequate training, inadequate inspections, non-enforcement of procedure, and failure to analyze risk. Some of the job factors identified were poor maintenance, use of malfunctioning equipment, inadequate corrective action, inadequate training, and poor inspections and corrections.

Once these basic causes have been identified, the necessary control steps can be instituted to break the sequence of events.

Dan Petersen summarizes these arguments by saying:

> Perhaps, however, our interpretation of the domino theory has been too narrow. For instance, using the investigating procedures of today, when we identify an act and/or condition that "caused" an accident, how many other causes are we leaving unmentioned? When we removed the unsafe conditions that we identified in our inspection, have we really dealt with *the cause of a potential accident?* (p17)

Conclusion

Accidental losses are caused by the failure to assess and control the risk. This then leads to the lack of, or a breakdown in, the safety management control system, which then creates basic causes, in the form of personal and job factors.

These basic causes are the very reason that unsafe conditions exist and that unsafe acts are executed, condoned, or tolerated.

Basic causes are the real problem in the cause and effect of accidental loss. Unsafe acts and conditions are only the symptoms.

CHAPTER FOUR

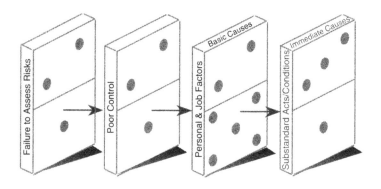

UNSAFE ACTS AND CONDITIONS

The fourth domino in the CECAL model is the domino representing unsafe acts and unsafe conditions , commonly referred to as the immediate causes of accidents.

Bird and Germain (1992) refer to them as the *sub-standard practices and sub-standard conditions*. They further explain the immediate causes of accidents thus:

> The immediate causes of accidents are circumstances that immediately preceded the contact. They usually can be seen or sensed. Frequently they are called "unsafe acts" (behaviors which could permit the occurrence of an accident) and "unsafe conditions" (circumstances which could permit the occurrence of an accident). Modern managers tend to think a bit broader, and more professionally, in *terms of sub-standard practices and sub-standard conditions* (deviations from an excepted standard or practice)(p.26).

Definition

An *unsafe act* is *the behavior or activity of a person that deviates from normal accepted safe procedure*. An *unsafe condition* is *a hazard or the unsafe mechanical or physical environment*. Bird and Germain list 16 substandard practices and 13 substandard conditions. They are as follows:

Substandard Practices

1. operating equipment without authority

2. failure to warn

3. failure to secure

4. operating at improper speed

5. making safety devices inoperable

6. removing safety devices

7. using defective equipment

8. using equipment improperly

9. failing to use personal protective equipment properly

10. improper loading

11. improper placement

12. improper lifting

13. improper positioning for task

14. servicing equipment in operation

15. horseplay

16. under influence of alcohol and/or drugs

Substandard Conditions

1. inadequate guards or barriers

2. inadequate or improper protective equipment

3. defective tools, equipment or materials

4. congestion or restricted actions

5. inadequate warning systems

6. fire and explosion hazards

7. poor housekeeping; disorderly workplace

8. hazardous environmental conditions

9. noise exposures

10. radiation exposures

11. high or low temperature exposures

12. inadequate or excessive illumination

13. inadequate ventilation (p.27)

Example

In investigating an incident (undesired event with high potential for loss under different circumstances), Al Rios (1998), identified three unsafe conditions and three unsafe acts that led to the situation that could have caused serious injury.

The unsafe conditions were:

1. The ground had shifted due to natural factors (underground mining).

2. The ground had not been supported for 11 hours.

3. Outdated procedures.

The unsafe acts that contributed to this incident were:

1. Timber supports had not been placed in position.

2. Written guidelines issued nine months previously had not been followed.

3. The three supervisors had all failed to warn the employee. (April 15, 1997)

Most Obvious

The unsafe acts and unsafe conditions are normally the most obvious events preceding the exchange of energy, and consequently, people tend to focus on them as being the true cause of accidents. The fact that unsafe behavior of an employee is involved also attracts much attention and as a result of extensively misquoted statistics, unsafe behaviors seem to be the main focus of accident-prevention programs.

Jim Howe, CSP, (1998) says:

> The claim that 90% (or similar fraction) of injuries are due to unsafe
> acts is a repetition of Heinrich's "research." Heinrich's conclusion was
> based on poorly investigated supervisory accident reports, which then, as
> now, blamed injuries on workers. He concluded that 88 percent of all
> industrial accidents were primarily caused by unsafe acts. Companies that
> sell behavior-based safety programs continue to mislead clients by
> perpetuating this folklore. (p.20)

Bird and Germain (1992) support this statement by saying:

> Also, an increasing number of safety leaders confirmed the results from
> research in quality control that 80 percent of the mistakes (sub-
> standard/unsafe acts), that people make are the result of factors over which
> only management has control. This significant finding gives a completely
> new direction of control to the long-held concept that 85 – 96 percent of
> accidents result from the unsafe acts or faults of people. This new direction
> of thinking encourages the progressive manager to think in terms of how
> the management system influences human behavior rather than just on the
> unsafe acts of people (p.26).

As Jim Howe, CSP, (1998) says, this conclusion that 88% of accidents are
primarily caused by unsafe acts has been misleading and has led management and
others to believe that unsafe acts were the main item to concentrate on to eliminate
accidents. In studying the research method that Heinrich, the National Safety
Council and others used to derive these statistics, it is of vital importance to note
that the principal of multiple causes was totally disregarded when these figures
were obtained. Heinrich (1959) said, "But *in no case* were *both* personal and
mechanical causes charged." (p.21) What the researchers did was mention only the
cause of *major importance*. This swayed the statistics. Who was the judge deciding
which cause, either immediate or basic, was of "*major* importance?" A description
of the methodology used is evidence that the statement that 88% of all accidents
are caused by unsafe acts is, in fact, as Howe puts it "folklore."

Heinrich (1959) describes the method used as follows:

> This difference (15%) added to the 73% of causes that are obviously of
> a man-failure nature, gives a total of 88% of all industrial accidents that are
> caused primarily by the unsafe acts of persons. Check analyses, made
> subsequently on a smaller scale, produce approximately the same ratios.

In this research major responsibility for each accident was assigned *either* to the unsafe act of a person or to an unsafe mechanical condition, but *in no case* were *both* personal and mechanical causes charged.

In addition to the research that resulted in the development of the above ratios, other studies have been made, one of chief interest being that conducted by the NSC. This showed unsafe acts for 87% of the cases and mechanical causes for 78%. An analysis made in 1955 of cases reported by the state of Pennsylvania showed an unsafe act for 82.6% and a mechanical cause for approximately 89% *of all accidents*. One reason for the difference in the number of accidents charged to personal or mechanical causes in the three studies described above is that, in the last two, the method permitted both kinds of causes to be assigned for the same accident, whereas in the study first mentioned only the cause of *major importance* was assigned.

Admittedly, judgment must be used in selecting the major cause when a mechanical hazard and an unsafe act both contribute to accident occurrence. (p.21)

The "White Paper" survey carried out by *Industrial Safety and Health News* (December, 1998), also revealed that 71% of managers surveyed believed that careless employee actions (unsafe acts) caused many accidents. (p.22)

Amazingly, the safety pioneers' method of statistical research has created one of the major safety paradigms of our times.

According to Fred A. Manuele (1997), the advice given by Heinrich's causation model has been wrongly focused.

Safety practitioners have prominently used Heinrich's causation model. Other causation models are extensions of it. However, the wrong advice is given when such models and incident analysis systems focused primarily on: characteristics of the individual; unsafe acts being the primary cause of incidents; and measurements devised to correct "man failure," mainly to affect an individual's behavior.

Heinrich also wrote, "… a total of 88% of all industrial accidents … are caused primarily by the unsafe acts of persons." Those who continue to promote the idea that 88 or 90 or 92% of all industrial accidents are caused primarily by the unsafe acts of persons do the world a disservice. Investigations that properly delved into causation factors prove this premise invalid. (p.30)

Heinrich et al. (1969), gives Dr. Zabetakis's theory of the unplanned transfer or release of energy that causes personal injury and property damage. They quote his explanation as follows:

> Most accidents are actually caused by the unplanned or unwanted release of excessive amounts of energy (mechanical, electrical, chemical, thermal, ionizing radiation) or of hazardous materials (such as carbon monoxide, carbon dioxide, hydrogen sulfide, methane and water). However with few exceptions, these releases are in turn caused by unsafe acts and unsafe conditions. That is, an unsafe act or an unsafe condition may trigger the release of large amounts of energy, which in turn cause the accident. (p.32)

Vital Factor

It is obvious that the unsafe environmental conditions and unsafe acts of people (substandard acts, high-risk behavior, or at-risk behavior as they are sometimes called) are always present in the causation of accidental loss. As the CECAL model shows, they are a factor in the string of events that lead to accidental loss. If they were removed, the contact and subsequent loss would not occur. Early safety philosophers relied entirely on attempting to remove the unsafe act and unsafe condition and, due to finger-pointing exercises, tended to (and still do), focus on the unsafe behavior of the individual without really delving into the reason for that person's unsafe behavior.

Heinrich et al. (1969) calls the immediate causes the symptoms:

> These are the factors in the accident/incident sequence that have historically been called the most important ones to attack. They are also the factors that receive the bulk of attention in governmental safety and health inspections around the world. (p.26)

As quite rightly stated, most safety legislation focuses on safe physical conditions and safe behaviors of employees rather than focusing on management systems, controls, and checks and balances that are integrated into the day-to-day management to ensure that the conditions are maintained to an accepted safe level and that behaviors are controlled.

Safety Paradigm

Having read numerous books, articles, and newsletters on current safety philosophies, we still appear to believe that the majority of accidental losses are caused by the unsafe acts of people. This idea was quoted extensively in Heinrich's book *Industrial Accident Prevention*, which was first published in 1929. In the fifth edition, which was co-authored by Dan Petersen, PE, CSP, and Nester Ross, D.B.A., the 10 original axioms on industrial safety are stated. Axiom 2 reads, "The unsafe acts of persons are responsible for a majority of accidents." (p.21)

88%, 10%, 2%

The first edition of Heinrich's book stated that 88% of all accidents were caused by unsafe acts, 10% were caused by unsafe conditions and 2% were attributed to acts of providence. The National Occupational Safety Association (in their *Questions and Answers on Occupational Safety and Health* series of books) further promoted these statistics. In the *Advanced Questions and Answers Manual* (1986) question 9 asks:

What is the key to accident prevention?

Answer: If we were able to eliminate all unsafe acts and unsafe conditions, about 98% of all accidents would be prevented. Unsafe acts cause approximately 88% of all occupational accidents. Unsafe conditions cause approximately 10% of all occupational accidents. Acts of providence or natural phenomena cause approximately 2% of all occupational accidents. (p.6)

In subsequent publications such as the *Health and Safety Training, General Course Manual* (1994) the following statement appeared, "Human error/inefficiency – 88% i.e., the human factor; High risk conditions – 10% i.e., the engineering factor; Acts of nature – 2% i.e., the inevitable." (p.23)

Unsafe acts and conditions are further defined in paragraph 14.3 as *"Acts, which are willful or negligent or knowingly ignoring set standards will sooner or later result in an accident or incident."* (p.25)

In paragraph 10, *high-risk conditions* are defined as *"any variation from accepted safety standards, which may be the cause of incidents and/or accidents."* (p.21)

Du Pont seminars teach that 96% of all injuries are caused by unsafe acts, "Because unsafe acts cause 96% of all injuries ..." (p.1.6), and justify this statement by declaring:

> When we conducted a 10-year study of all serious injuries occurring at Du Pont sites – at offices, on refineries, in transportation, at all kinds of plant sites – we learned that 96% of our injuries were caused by the unsafe acts of people and poor work practices. (p.1.5)

The *Du Pont Management Safety Seminar Manual* contains their "iceberg concept." According to this diagram, the unsafe acts are at the bottom of the iceberg and they lead to first aid cases, medical treatment, and then the tip of the iceberg is reflected as lost-time injuries. In viewing the CECAL model, it is obvious that this focus is only concentrating on half of one domino in an entire sequence of events that cause the loss. Focusing on the unsafe acts alone will not necessarily mean the interruption of the loss causation sequence. Dr. Mark A. Friend (1997) asks the question: "Since 90% of all accidents and adverse incidents are due to human error, should safety efforts be focused on correcting these errors? The answer: False."

He then explains:

> Although 90% of all accidents and adverse incidents may, in fact, be caused by human error, safety efforts must also encompass engineering principles. When asked whether the engineering or human relation's school should be emphasized, Fred Manuele says, "A curse on both their houses." A seminar leader once suggested that since 95% of all safety problems are due to human error, 100% of solutions should address the human element. His reasoning: by taking such action, nearly all problems will be covered. Such thinking is naive and dangerous. Addressing the human element can solve some problems. But ignoring engineering solutions is akin to fixing the problem with a short-lived solution that will fail when the slightest change occurs. (p.36)

Why?

Safety practitioners, management, and union management as well as fellow employees have asked this question a thousand times, "Why do people commit unsafe acts?" There is no simple answer to this question and each accident investigation that has involved unsafe acts has always indicated that there were reasons and motivation that caused the person to commit the unsafe acts.

Dan Petersen (1998) attempts to give an explanation by showing how workers personally benefit from working unsafely. He gives four reasons:

1.   The advantages and satisfaction to be gained by the worker at the particular moment seems greater to him than the disadvantages and dissatisfactions.

2.   The unsafe act "makes real sense" to the employee. If he is challenged, he will explain to the foreman exactly why he thinks his way is the most sensible way to do the job. Typically, the older employee will justify himself by saying that he has been doing it that way for years.

3.   The unsafe act actually gives the worker personal satisfaction; it may attract the attention of co-workers; gain their approval and admiration; give him either the thrill of taking a chance or the satisfaction of bucking authority – or even paying back an imagined grudge; it may make him feel daring; and it may involve many other personal incentives.

4.   To the worker, his "unsafe" act may be perceived as having definite job related advantages – advantages that include either such monetary incentives as getting his job done sooner, thus increasing his work-output and his take-home pay – especially if he is on piece-work pay – or personal incentives such as avoiding extra effort or fatigue and having more "personal control" over product quality. (p.232)

Having worked with employees in industrial situations for a number of years, my personal conclusions are that people commit unsafe acts because they are (we are) human beings. When questioning motorists as to why they travel at 60 mph instead of the limit of 55 mph, they all generally agree that the advantages of the risk generally outweigh the disadvantages. They also rationalize that breaking the rules "a little bit" will never hurt anyone. I was once told that an African philosophy is that if you are milking the cow and take a sip of the milk, the cow will not die.

Having been involved in numerous accident investigations, I saw that the unsafe acts were not unique to the accident and had been executed many times previously. Unsafe conditions don't merely pop up and cause an accident but have also been there for a long while. Complacency and condoned practices are what cause unsafe acts and unsafe conditions. In numerous accidents, fellow workers, supervisors, and others ignored the unsafe act. Once the unsafe act resulted in a contact and an accidental loss such as injury, the employee was immediately confronted with his unsafe behavior. The same happens with unsafe conditions. Another safety paradigm is disciplining employees for committing unsafe acts. I asked the question, "How can an employee be disciplined for committing an unsafe act when one unsafe condition is present in the workplace?" Safety management is a two-edged sword and both sides should be viewed equally. Allowing unsafe acts to go unheeded is tantamount to committing an unsafe act and one deviation from accepted workplace standards is also totally unacceptable.

Example

In the 3C-accident investigation there were found to be numerous unsafe acts and conditions. There was a failure to warn, operating at unsafe speed, safety devices made inoperable, defective equipment, improper positioning, people not following procedures, defective equipment, absence of a warning system, and a hazardous environment. Model 4.1 shows the unsafe acts and conditions listed under the fourth domino.

Unsafe Acts:

- The controller was not checked.

- The machine was operated at an unsafe speed.

- Safety devices had been rendered inoperable.

- The operator was not aware of the danger of the faulty sensor.

- The computer indicated the wrong position.

Unsafe Conditions:

- The transformer was running too hot.

- Inadequate ventilation.

- The stop limit switch was not in place.

- The environment was hazardous and there was inadequate lighting.

- No warning system existed.

Model 4.1 - 3C-accident investigation.

Multiple Causes

The principle of *multiple causes* states that *accidents and other problems are seldom, if ever, the result of a single cause.* This means that there is always more than one cause for an accident and seldom will an accident be as a result of a single isolated unsafe act or unsafe condition. Normally, there are both acts and conditions. If one counts only the unsafe acts, one could easily derive the statistic that the majority of accidents are caused by the unsafe acts.

Research (Company XYZ)

During the financial years of 1996, 1997, and 1998 an analysis was made of injury reports as to what percentage of these accidents were caused by unsafe acts and what percentage by unsafe conditions. The principle cause of the accident was also used to derive these statistics.

Findings

During the 1996 period a total of 352 injuries was recorded, with 208 being the result of the unsafe act and 131 as a result of the unsafe condition. This meant that some 40% of the accidents were caused by the unsafe conditions and 60% by the unsafe acts. Model 4.2 shows the analysis produced.

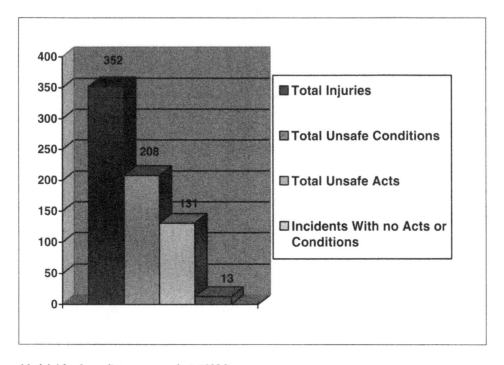

Model 4.2 – Immediate cause analysis (1996).

In 1997, the picture changed somewhat with 388 total injuries reported. One hundred and thirty were the result of unsafe acts and 180 the result of unsafe conditions. Rounding off the figures, some 34% of accidents were caused by the unsafe act and 47% by the unsafe conditions. In the 1997 figures, 78 injuries had neither acts or conditions listed and were therefore removed from the equation.

1997				
Total Injuries and Incidents	**Total Unsafe Acts**	**Percentage of Unsafe Acts**	**Total Unsafe Conditions**	**Percentage of Conditions**
388	130	33.51%	180	46.40%

Model 4.3 – Immediate cause analysis (1997).

During the year 1998, the same study was undertaken and the findings were as follows: There were 341 injuries recorded as a result of accidents, 114 were as a result of the unsafe acts and 172 as a result of the unsafe conditions. These then represented 34% caused by unsafe acts and 50% by unsafe conditions. Model 4.4 shows this research.

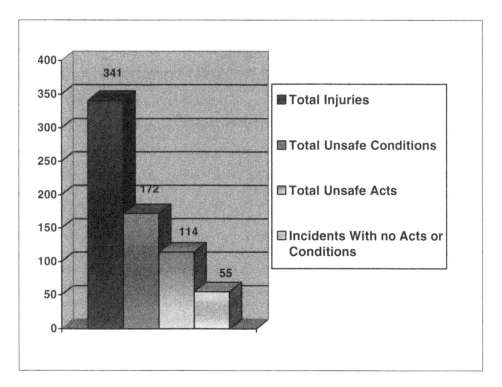

Model 4.4 – Immediate cause analysis (1998).

Unless a company has conducted research as to what are the immediate causes of their loss-producing accidents, it would be unwise to quote generic statistics, which are slanted toward either the unsafe act or the unsafe condition. It is also opportune to acknowledge that unsafe acts and unsafe conditions are factors in the cause and effect of accidental loss. The percentages should be compiled by accurate data collection and statistical analysis, and be pertinent to a particular workplace. Generalizing these statistics and quoting unsafe acts as the ultimate and only cause of accidents, wrongly points the finger at the worker who is only a pawn in the safety management game. The injured employee is only a victim of a failure of the management system.

Basic Causes

Unsafe acts and unsafe conditions exist or are caused by underlying causes in the form of personal and job factors. Rather than treat the symptoms of a failure of the management system, these immediate causes should be used as a means of probing, identifying, and rectifying the real problems in the form of personal and job factors. The work environment impacts heavily on the behaviors of people and can dictate how they act. In discussions with Frank E. Bird Jr. he once told me, "If a manager cleans up the work environment, he also cleans up the thought processes of the people in that environment." There are reasons for unsafe acts and there are also reasons for unsafe conditions. The best way to eliminate the immediate causes is to find and eliminate the basic causes.

Example

In Case Study 4 (*Questions and Answers on Effective Accident/Incident Investigation* 1990), the injured employee's recollections of his actions, which led to devastating personal consequences for him, are related as follows:

> Then came the most terrible day in my life, 21st September 1980, at 2.30 pm. I was working on the steering clutches of a piece of heavy earth-moving equipment. To replace core lead inserts on a steel band I had to grind off some old welding. I was working on a rough and ready construction site and the only grinder (angle grinder) had been bolted onto a box and the box had been filled with rocks for added "safety" and stability as a base.

I could not reach inside the break bands to grind with the guard in position. Without thinking, I removed the guard. I did not realize that I was adding more danger to an already hazardous situation. Suddenly, as I commenced working, the wheel disintegrated. I felt a terrible pain in my stomach. A wedge-shaped piece of the wheel sliced its way through my abdominal muscles and into my abdominal cavity. It caused a 12-cm long wound in my abdominal muscles. It then left my body through an 8-cm gash in my back and smashed into my lower leg .... (p.51)

This accident left the employee on the construction site for some two hours and he only reached the hospital two hours later. Four hours and 10 minutes after that, the surgeon arrived and he was in the operating theater for three hours and 30 minutes. The employee's foot bones were smashed and his foot had to be amputated. He then spent a further 28 days in hospital and has had tremendous trouble with his artificial limb and doctors even suggested a second amputation at the knee.

Behavioral Based Safety

Dr. Thomas R. Krause (1997) defines behavioral based safety as:

The phrase *behavioral based safety* refers strictly to use of applied behavior analysis methods to achieve continuous improvement in safety performance. These methods include identifying operationally defining critical safety-related behaviors, observing to gather data on the frequency of those behaviors, providing feedback, and using the gathered data for continuous improvement. (p.30)

Dr. Krause continues by saying:

Wherever it appears in this book, the statement that most accidents are "caused by" at-risk behavior, the type of cause referred to is known as the final common pathway. Those critical behaviors are "produced" by management systems of the site and therefore decreasing them is the key to accident prevention. (p.30)

In discussing behavioral analysis he continues by saying:

Behavioral analysis is one of the basic tools of the employee-driven safety process. Most safety problems involve someone's behavior. Behavior is controlled by antecedents (things that trigger behavior), and by consequences (the things that both follow from behavior and influencing it). (p.30)

Using a procedure based on safety observations, Krause quotes a case study of outcomes at a paper mill employing 425 employees. The case study is summarized by his quotation, "The mill realized a 47% reduction in injuries in the first 2 years of their process." (p.50) He concluded that by focusing on behaviors critical to safety performance, this papermill achieved a steady decrease in reportable injuries. He further stated that, "Ongoing peer to peer observations and feedback about safety performance has become part of the mill's culture: it's what the employees talk about." (p.50)

Dr. Krause also motivates behavior-based safety by saying:

> The reason to focus on behavior is that when an incident occurs, behavior is the crucial final common pathway that brings other factors together in an adverse outcome. Therefore, ongoing, upstream measurement of the sheer mass of these critical at-risk behaviors provides the most significant indicator of workplace safety. (p.19)

He describes the behavior-based safety improvement process as:

- identify critical behaviors
- identify root causes
- generate potential actions
- evaluate possible actions
- develop action plans
- implement action items (pp.79-89)

His safety improvement process model indicates that if the problem is not solved, the identification of critical behaviors must be reverted to and if the action items are complete, reverts to identifying root causes and generating the potential actions. If the actions are incomplete, re-implement the action items.

The success rates in the paper mill case study discussed by Krause measure the recordable injury rate decrease and the number of employees injured. If measuring the degree of injury is accepted as a good indication of safety, then the results of behavioral-based safety according to Dr. Krause are significant. It is often been quoted however, that the measurement of injury, and specifically a certain degree of injury, is not reliable, as the injury is largely dependent on the luck factors 1, 2 and 3 as will be discussed in chapters 5, 9, and 10 of this publication.

Safety Success

The late James Tye, who was the director-general of the British Safety Council was once quoted as saying that (injury) frequency rates were measurements of failure, rather than success. He further explained that since the under reporting of accidents and injuries could be as high as 80% in some U.S. industries, he concluded that frequency rates were irrelevant in measuring safety "performance." According to Bird and Germain, measurement of management control is a far better and reliable measurement of safety performance. Bird and Germain also quote Charles E. Gillmore, who, during an address at the National Safety Congress, said, "What is the sense in measuring if the loss must occur before you can act? That's reaction, not control." (p.49) Is Dr. Krause measuring the success of the behavior-based safety systems merely by measuring the degree of injury rather than degree of control?

At-Risk Behavior

Dr. E. Scott Geller (1996) also quotes Heinrich:

> Heinrich's well-known law of safety implicates at-risk behavior as the root cause of most near-hits and injuries. (p.76)

> Over the past 20 years, various behavior-based research studies have verified this aspect of Heinrich's Law by systematically evaluating the impact of interventions designed to lower employees' at-risk behavior. At-risk behaviors are presumed to be a major cause of a series of progressive, more serious incidents. (p.76)

Geller refers to "near-hits" but the American Society of Safety Engineers' technical dictionary does not define the concept "near-hit" but does define the concept near-miss.

Based on this approach of presuming unsafe acts to be the major cause of serious incidents (accidents) as stated by Geller, Jim Howe, CSP (1998), says that:

> The behavioral safety approach is biased. It ignores hazards and risks and focuses on "critical worker behaviors," which would permit working in a hazardous environment. This almost always leads to the

implementation of low level controls, safety procedures and personal protective equipment instead of more effective engineering control. (p.20)

Geller (1996) also warns that safe acts should not be ignored and that corrective feedback should be used.

> Since adverse behaviors contribute to most if not all injuries, a Total Safety Culture requires a decrease in at-risk behaviors. Organizations have attempted to do this by targeting at-risk acts, exclusive of safe acts, and using corrective feedback, reprimands, or disciplinary action to motivate behavior change (p.86).

Differences

Jim Howe, CSP, (1998), defines the difference between a systematic approach and the behavioral systems approach as follows:

> There is a substantial difference between a systematic approach to workplace health and safety and a behavioral safety approach. The system approach takes an objective and unbiased view of the workplace by:
>
> 1. Identifying hazards.
> 2. Estimating the level of risk for each hazard.
> 3. Controlling hazards according to the hierarchy.
>
> Behavioral-based safety programs appeal to many companies because they make health and safety seem simple, do not require management change, focus on workers, and seem cheaper than correcting health and safety hazards. (p.20)

Acts and Conditions

Unsafe acts and unsafe conditions are factors in the accident sequence. Focusing on only one factor in a multiple causation sequence would be ineffective and therefore a holistic control approach is required.

The unsafe acts and unsafe conditions will continue and will always be the event that result in an exchange of energy, which in turn causes the loss.

The unsafe acts and unsafe conditions will remain the obvious causes of the accidents but are clearly only the symptoms of other greater failings within the organizational system. To have a balanced approach, both employees, supervision, and management have specific duties in safety management.

They are, according to McKinnon (1995):

Management

Management is ultimately responsible for safety and health and their responsibility is to:

- Set objectives and develop a policy for safety and health.
- Bring operations in line to comply with applicable legislation.
- Delegate responsibility and authority to those at various levels in the organization for certain parts in the loss control program.
- Ensure that safety and health information is an integral part of training and operations.
- Ensure that contractors comply fully with company and other applicable safety regulations.
- Maintain an industrial hygiene monitoring system.
- Set a good example by attending safety and health meetings and taking action on accident investigation reports.

Various other levels of supervision also have specific roles to play in safety management. The employees' roles are as follows:

- Must work according to safety and health standards and procedures.
- Must observe safety and health rules and regulations.
- Must report hazardous conditions and unsafe practices.
- Develop and practice good habits of hygiene and housekeeping.
- Must use personal protective equipment properly and maintain it in good order.
- Report all injuries and incidents.
- Assist in developing safe work procedures.
- Make suggestions for improving safety and health conditions and procedures (*Highlight 46*, p.2).

Conclusion

The failure to assess and control risks creates a breakdown in the management control function, which creates personal and job factors, which in turn lead to unsafe acts and unsafe conditions. Unsafe acts and unsafe conditions are the symptoms of the loss-causing situation and should be used to identify and eradicate the basic causes. An unsafe act or an unsafe condition will inevitably lead to a contact with a source of energy, which will lead to loss.

CHAPTER FIVE

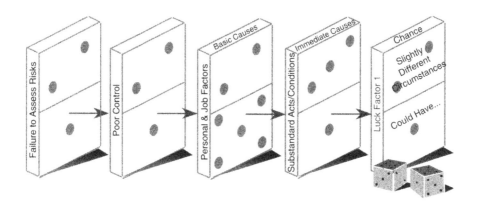

LUCK FACTOR 1

We have already discussed how failure to assess and control the risk triggers off poor controls in the form of an inadequate safety system. This causes personal and job factors. These basic accident causes create a climate that breeds unsafe behaviors and unsafe conditions. The stage has been set and invariably the next event is either a contact with the source of energy or a near-miss, depending on Luck Factor 1.

It would be appropriate at this stage to redefine the terms *accident* and *incident*.

Accident

"An accident is an undesired event, which causes harm to people, damage to property or loss to process" according to Bird and Germain (1992). (p.18)

Incident (Near-Miss)

Bird and Germain (1992), define an incident as "an undesired event, which, under slightly different circumstances could have resulted in harm to people, damage to property or loss to process" (p.19).

It is therefore concluded that there is no logical explanation why some unsafe acts end up as accidents and the same unsafe act, under slightly different circumstances, ends up as an incident. The difference is determined by luck.

Luck Factor 1

In referring to the CECAL model, it is proposed that Luck Factor 1 determines the outcome of an unsafe act, unsafe condition, or combination thereof. The outcome can either be contact with a source of energy or merely an incident. An incident is an undesired event, which, under slightly different circumstances, could have caused a loss. It has potential to cause a loss.

Mark Rothstein (1998) puts it this way, "On the other hand, the absence of an accident doesn't mean there was no violation – it may only reflect the employer's good fortune." (p.135)

Measurement

Dan Petersen (1998) talks on safety measuring systems as follows:

> In the early days of safety, accident measures, such as the number of accidents (injuries), frequency rates, severity rates, and dollar costs, were used to measure progress. Although it became clear long ago that these measures offer little help, they continue to be used today. Why should we consider other measures? "Results" measures nearly always measure only luck – unless they concern a huge corporation that generates thousands of injuries. (p.37)

Energy Source

What differentiates an accident from an incident is *contact* with a source of energy or substance greater than the threshold limit of the body. This means that there was an exchange of energy that caused the loss. The contact with this source of energy is always as a result of an unsafe act or unsafe condition. In thousands of documented cases, unsafe acts have been committed and there has been no contact or no consequence whatsoever. Similarly, the same unsafe acts have been committed and, under slightly different circumstances, have resulted in contact with the source of energy and a subsequent loss to either people, property, or the process. The only plausible explanation is that Luck Factor 1 determines the outcome of unsafe acts and conditions.

Vaal Reefs

My first safety training in the USA was conducted at a large mine. While I was introducing myself to the group, a young miner in the audience stood up and said to me, "Hey you, Crocodile Dundee, what the heck can you teach us about safety?" Being somewhat taken aback by this statement I first of all inquired what he meant by Crocodile Dundee. On hearing my accent he thought I was from Australia and not South Africa. I then asked him what he meant by his statement. He stood up and continued, "You 'fellas' in South Africa have just killed 105 people at a mine where a 12-ton locomotive fell down the shaft and collided with a man cage. It took it down to the bottom crushing it completely and killing all 105 occupants." There was a terrible silence in the room and he continued, "If those types of accidents happen in South Africa, what can YOU teach us about safety?"

He was absolutely correct. On May 10 1995, at the Vaal Reefs No. 2 main shaft, a 12-ton locomotive pulling a man carriage fell down the shaft. At the same time a fully loaded man-cage was moving slowly down the shaft. At approximately 50 meters below the '56' level the collision took place and the man-cage plummeted 500 meters to the shaft bottom, where it was crushed by the locomotive. The 7-meter high, two-deck cage was crushed down to 1.5 meter. *The Sunday Times* (1995) report entitled "The level 72 horror" tells the story:

> A free falling 12-ton locomotive glanced against the lift and pieces of metal smashed through its top, hitting Mr. Quluba on the head. His neck was fractured and one metal piece caused a fast bleeding wound just above his forehead. Luckily he survived and the lift made its way to the top.

> But 105 other miners were not so lucky. They died when the locomotive and its carriage smashed into their cage on level '56', sending it plummeting to the bottom at 120 Km/h (70 mph). (p.9)

On the same page, Peter DeIonno continued with the story:

> Evidence will be presented to the disaster inquiry that barriers intended to stop underground trains from approaching the shaft mouth were often

removed at the mine to speed up handling of rail carriages. In free fall, the train took 3 seconds to catch up 50 meters down with a fully loaded two deck man-cage lift that had just passed downwards at 16 meters a second. The National Union of Mineworkers has declared May 17 a day of mourning and is planning marches to protest against poor safety standards at mines. (p.9)

The young miner certainly had a point and had every reason to question my ability. I then addressed the group, which consisted mainly of underground miners with an average of 17 years' experience. I asked them if a locomotive had ever fallen down one of their mineshafts. There was an uncomfortable silence in the room, which was eventually broken by an old timer who stood up and said, "Yes, in 1960 we dropped an 8-ton locomotive down the main shaft." I asked him how many people had died as a result of this accident. He replied, "None, nobody was injured." I asked, "Why?" and was greeted by silence. I then repeated my question. "At Vaal Reefs a loco fell down a shaft and 105 people died. Here, in 1960, a loco fell down a shaft, the same as at Vaal Reefs, but no one died and nobody was even injured, why?" There was a silence in the room and one of the miners stood up and said, "We were lucky." How right he was. They were lucky that the cage was *above* the level from which their locomotive plummeted. The Vaal Reefs scenario was clearly an example of bad luck, the cage having being positioned *below* the level from which the locomotive fell.

Examples of Luck Factor 1

Bearing the Vaal Reefs level 72 disaster in mind, certain research was carried out at an underground mine in an effort to prove that Luck Factor 1 determines the outcome of an unsafe act or an unsafe condition. This particular mine had never had a disaster similar to Vaal Reefs but upon investigation found that it had had three opportunities for a similar disaster.

Accident 1

In this particular accident in 1989, the report stated "The rail came down hitting the end of the transporter, knocking them over the scotching chains and down the shaft..." There were no injuries as a result of this 1,000-lb. vehicle falling down the mine shaft. (December 28, 1989)

In 1994, another vehicle fell down the shaft. "As tension was added by the hoist, the cage broke loose and sprang up 6 to 8 feet, overturning the battery onto the station and the flat (car) fell down the shaft." (April 11, 1994)

A similar accident occurred in 1995, "This caused the leading flat (flat car) to tip up, come uncoupled and roll into the no. 2 compartment of the shaft. This occurred very quickly and was over in a matter of only a few seconds." (March 20, 1995)

The above three accidents did cause property damage and minor business disruption but there were no injuries and no fatalities primarily due to Luck Factor 1. The circumstances proved to be fortunate because none of the items fell on a cage full of miners.

Warnings

S.L. Smith (1994) says that, "If enough near-misses occur, the question is not, "will an actual accident ever happen," but "when will it happen." (p.33)

Many call the incidents "warnings" and quite rightly so. Safety management is perhaps the only management science where numerous warnings are given before a contact and loss occurs. The only reason that a contact occurs, or does not occur, is because of the luck factor. The warnings should be heeded nevertheless.

Accident Investigation

In investigating an accident in which a worker climbed a poorly supported electric utility pole and was killed when both he and the pole fell to the ground, it seemed a clear-cut case that the man had committed an unsafe act. Thorough investigation and some five days of gathering further evidence showed that a fellow worker had been doing exactly the same as the victim except on a different pole. This pole was half a mile distant from the fatality site. Investigation showed that this pole was also not completely compacted. The worker had ascended the pole on numerous occasions and had completed the running of the conductors through the isolators. When trying to analyze why the one pole fell, killing its climber, and the other pole did not, one can only conclude that the electrician on the distant pole was lucky.

Near-Miss Reports

In reviewing monthly near-miss reports, the luck factor is very apparent. Near-miss no. 4 was as follows: "Contractor did not lockout the disconnect (switch) before starting the job, electricians almost turned power on but checked first to see if someone was working." (September 1, 1998)

This near-miss indicates that the contractor was lucky that the electricians decided to check first to see if somebody was working on the line.

The next near-miss not only indicates the luck factor but also the high potential for loss. Near-miss no. 2 is as follows:

> We were trying to open the man-way doors, to go to the surface. We could not open the door because the door to the shaft and the door to the station were open. We finally got the door open and the pressure of the air pulled me in toward the shaft. I almost ended up falling down the shaft had it not been for the wire (September 2, 1998).

Once again, the luck factor determined the difference between a contact and no contact and in this case, circumstances were slightly different and the man was not sucked into the shaft.

Accident Ratios

There have been numerous accident ratios proposed over the years. Heinrich (1959) first proposed the concept that there were some 300 accidents to every serious injury. The Frank Bird Ratio of 1966 expanded on this theory and showed 600 incidents with no visible sign of loss for every serious injury. The Tye/Pierson Ratio mentioned 400 undesired events for every serious injury. The main question is what determines the difference in outcomes of these undesired events? What differentiates between contact and non-contact? The only logical answer is the luck factor.

Tarrants (1980) summarizes the luck factor by saying: "The study of errors and near accidents usually reveals all those situations that result in accidents plus many situations that *could* potentially result in accidents but that have not yet done so. (p.117)

Summary

There have been accident ratios calculated in the past and there will be new accident ratios calculated in the future. In generalizing all accident ratios, it can be said that for every one serious injury there are *some* minor injuries, *more* property damage accidents and *plenty* of incidents which, under slightly different circumstances (luck, chance, fortune), *could have* resulted in accidents with loss.

Many organizations rely on this luck factor rather than on control for reducing the number and degree of their accidental losses. A near-miss reporting and assessing program is so vital to an organization as it determines what *could have* happened under slightly different circumstances. It is important to assess the potential, severity, and frequency of near-misses, as well as accidents that produce a loss. In discussions with management who venomously defends its lack of or few injury accidents, I always ask the question, "Are you in control or relying on luck?"

CHAPTER SIX

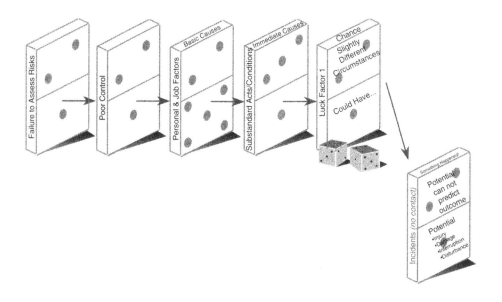

UNDER SLIGHTLY DIFFERENT CIRCUMSTANCES

So far, the cause and effect of accidental loss has shown us that failure to assess and control the risks creates basic causes that cause an environment that promotes unsafe acts and unsafe conditions. These, as determined by Luck Factor 1, either result in a contact (exchange of energy) or an incident (near-miss).

Warnings

These incidents have often been called near-misses, narrow escapes, and warnings. Irrespective of their terminology, they are the real "safety in the shadows" according to S.L. Smith (1994). (P.33)

We have been aware of incidents, near-misses, and warnings for a number of years. Safety pioneer Heinrich (1959) introduced his 10 axioms of industrial safety back in the 1930s. His third axiom reads as follows:

> The person who suffers a disabling injury caused by an unsafe act, in the average case, had over 300 narrow escapes from serious injury as a result of committing the very same unsafe act. Likewise, people are

exposed to mechanical hazards hundreds of times before they suffer injury. (p. 21)

Misuse of Terminology Creates Confusion

As a trainee safety adviser I was taught that an accident resulted in a loss and that an *incident* was a *near-miss, warning, or narrow escape.* Since then, confusion has been created by all events being grouped under the same heading of *incident.* We had an undesired event at the office when the secretary dropped the coffeepot, which shattered, spraying hot coffee all over the rug. This was referred to as the "coffee incident." On the very same day, the newspaper reported there had been a malfunction at a nuclear power station, which was also referred to as an *incident.* To avoid confusion, it is recommended that we return to basics and redefine the concepts. Bird and Germain (1992) define an *accident* as *"an undesired event that results in harm to people, damage to property or loss to process." (p.18)* They defined an *incident* as *"an undesired event, which under slightly different circumstances, could have resulted in harm to people, damage to property, or loss to process."* (p.20) It is interesting to note that this is the first half of their definition of an incident.

Krause (1997) defines an *accident* as *"the unplanned result of a behavior that is very likely a part of the culture on the site."* (p.292) Although his definition comes from a behavioral safety aspect it should be noted that the definition implies all accidents are the result of behavior and no mention is made of an unsafe work environment. He defines an *incident* as *"the unplanned result of a behavior that happens not to cause injury or damage."* (p.292) Both definitions therefore clearly indicate that: An accident causes some form of loss and an incident does not cause loss.

Further Refinement

A further refinement has been made in distinguishing the difference between an incident and a near-miss. It should be remembered that these definitions define the narrow difference between the incident and the near-miss.

An *incident* is any exchange of energy with contact that under slightly different circumstances could have resulted in injury, property damage, or loss of process. A new element has been added to this definition in that there was an exchange of energy and contact but no loss.

A *near-miss* is redefined as any incident where an exchange of energy occurs, but there is no contact and which under slightly different circumstances could have resulted in injury, property damage, or loss of process. So here we have exchange of energy, contact, but no loss. The near-miss is distinguished by the concepts of an exchange of energy, but no contact or loss.

Unsafe Acts and Conditions

Many people are not sure of the difference between an unsafe act, an unsafe condition, an incident, or a near-miss. An *unsafe act* or *condition* is defined as *a situation or act that has potential for loss*. The word potential means the probable event is imminent but no actions have yet taken place. The minute there is an event, that unsafe act or condition then becomes an incident or a near-miss. Both definitions call for the presence of energy, whereas an unsafe act or unsafe condition has not yet created an energy flow.

The following case studies are used to explain the differences between accident, incident, and near-miss.

Example of an Accident

The employee was descending the stairway from the first floor office when his foot slipped on the second step from the bottom, causing him to fall. His elbow hit the bottom of the stair, causing bad bruising and a hairline fracture to the elbow.

Here we have an accident, which has resulted in loss in the form of an injury.

Example of an Incident

In the village of Ipswich in England, a 2-year-old child, Jamie Dighton, fell 28 feet from a second-floor balcony onto the ground below. Immediately after landing he stood up and only had two small bruises to show for his experience. X-rays were taken but they showed he had suffered no injury whatsoever. Here an undesired event had taken place; there was an exchange of energy but no loss. This was an incident.

Example of a Near-Miss

An example of a near-miss is the following case study. A man was moving a heavy electric motor off of its mountings using a crowbar to slide the motor from its position. In tensioning the bar it suddenly flew out of his hands, just missing the face of his colleague by a few inches. His colleague said that he felt the wind on his cheek as the flying bar flew through the air narrowly missing him.

This event clearly fits the description of a near-miss energy exchange, but no contact.

Simplicity

Although it is interesting to examine in finer detail the differences between an incident and a near-miss as defined above, these two concepts will be grouped together for simplicity to explain the concept of incidents, incident reporting, and its importance in the cause, effect, and control of accidental loss sequence (CECAL).

Ratios

Heinrich was perhaps the first author to document the fact that for every serious injury there had been a number of incidents and near-misses that could have given warning of the injury. His ratio, reproduced as Model 6.1, proposed that for every 300 accidents there were 29 events that caused minor injury and one that caused disabling injury.

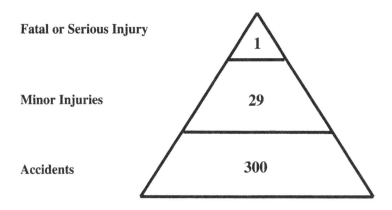

Model 6.1 - The Heinrich Accident Ratio (McGraw-Hill,1959).

In Model 6.2 the 1966 accident ratio study reproduced in Bird and Germain changed the concept by proposing that for every 600 incidents, 30 property damage events occurred, 10 events resulted in minor injury and one serious or major injury was experienced. This analysis was made of nearly 2 million accidents reported by approximately 300 participating companies employing 1.7 million employees. As Bird and Germain (1992), report on the process involved in the study:

> Part of the study involved 4,000 hrs of confidential interviews by trained supervisors on the occurrence of incidents that under slightly different circumstances could have resulted in injury or property damage. In referring to the 1:10:30:600 ratio, it should be remembered that this represents accidents and incidents reported and not the total number of accidents or incidents that actually occurred. (pp.20-21)

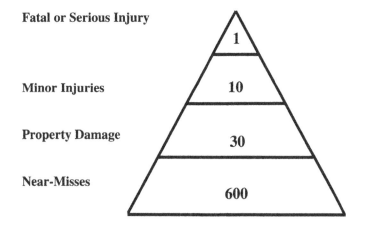

Model 6.2 - The Bird-Germain Accident Ratio.

Tye-Pierson Ratio

In the United Kingdom during the years of 1974 to 1975, a study was conducted on behalf of the British Safety Council. Based on a study of almost 1 million accidents in British industry, the Tye-Pierson Ratio was deduced. The study was concluded by stating that, "There are a great many more near-miss accidents than injury- or damage-producing ones, but little is generally known about these," (British Safety Council, *Safety Management*).

Model 6.3 reproduces the Tye-Pierson Accident Ratio (1974/75) showing that for every one serious injury there were three minor injuries, 50 first aid, 80 damage-causing accidents and 400 near-misses. (Model 6.3)

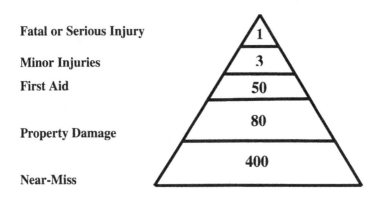

Fatal or Serious Injury   1

Minor Injuries   3

First Aid   50

Property Damage   80

Near-Miss   400

Model 6.3 - The Tye-Pierson ratio.

Health and Safety Executive

In 1993 the Health and Safety Executive of Great Britain came to a similar conclusion with its accident ratio. It proposed that for every serious or disabling injury 11 minor injuries occurred and in excess of 440 property damages were recorded.

Model 6.4 shows the HSE Ratio indicating for every one serious or disabling injury 11 minor injuries occurred and 441 property damage accidents occurred.

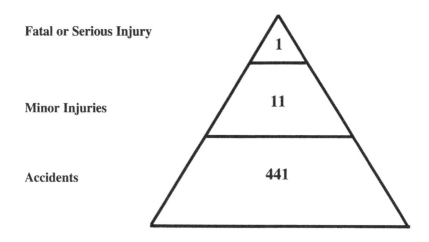

Model 6.4 - The HSE Ratio.

USA

A brief investigation as to the actual ratio of injuries to near-misses was conducted at a large industry in the USA and the following ratio was determined. For every one disabling injury there were 15 reportable injuries, four first aid cases, one property damage and 286 near-misses. This ratio clearly showed that there was a correlation between reportable injuries and disabling injuries but also indicated the lack of reporting of property damage accidents. The near-misses reported indicated a good near-miss program was in operation.

Numerous industries, governments, organizations, and safety practitioners use the number and severity of injuries as the only measurement of safety. According to James Tye, the late director of the British Safety Council, these are measurements of failure. Measurements of injury and severity, as well as the costs thereof, are all reactive measurements and are not predictive. They are largely fortuitous anyway.

Perhaps the most important statistic in any safety management program is the accident ratio showing the proportion of near-misses, property damages, minor injuries and serious injuries. Many authors and organizations have also compiled accident ratios but only between first aid cases, reportable injuries and serious injuries. What has been left out in many cases is the bulk of the iceberg, the hidden part, the number of near-misses or warnings that an organization gets before a major loss occurs.

South African Ratio

McKinnon (1992) in *Safety Management* magazine (September 1992), elaborated:

> A ratio such as this is truly invaluable, since it makes prediction possible. As soon as too many near-miss incidents are reported in a specific area, it is a clear sign to the safety staff that a more serious accident is about to occur and they can take preventive action.
>
> Bear in mind that the only difference between a near-miss accident and a serious accident is generally just a matter of luck. Take for example a brick falling off a platform and narrowly missing the head of a passerby.
>
> The accident occurred the minute the brick started to fall. The fact that it missed the passerby was not due to any safety institution but was purely fortuitous. While this is recorded as a near-miss, the implications are very serious.
>
> NOSA decided that statistics, which will enable them to construct a ratio, would be of great value to industry. "We want to determine whether there is a unique ratio in this country (South Africa) or whether Bird's Ratio of 1:10:30:600 was accurate." (p.12)

Trends

Writing for *Occupational Hazards*, S.L. Smith (1994) says:

> Near-miss investigations war with the tradition of using an accident to trigger a thorough look at safety conditions and training. Williams suggested that if the purpose of safety programs is to prevent accidents, then tracking near-misses offers companies a better opportunity to lever their preventive efforts. Near-misses can help firms pinpoint trouble areas and focus their safety training. (p.34)

Numerous organizations around the world have posters and banners depicting that their aim is to be injury free. Numerous management-training programs also focus on injury prevention and tend to be specifically aimed at recordable injuries. Attention is given to that injury that requires reporting to the board of directors, the local safety and health enforcement agencies or that injury that will ruin their "safety record."

If one examines all the ratios and the research that has proven that there are numerous near-misses before there is the occurrence of a severe injury, it can be deduced that it is impossible to be injury free unless you are accident free. To be accident free, one has to be incident free. To be incident free, there must be no unsafe acts and no unsafe conditions. In the real world, it is almost impossible to eliminate all unsafe acts and unsafe conditions, but the near-misses can give an indication of which unsafe acts and unsafe conditions have greater *potential* for causing the accident that in turn will cause the injury.

Frank E. Bird Jr. convinced me that the key to incidents was, "If you look after the incidents, the accidents will look after themselves." One has to reduce the base of the triangle before the injuries that made up its peak can be reduced. Focusing only on injuries treats symptoms and not problems. After all, what sank the *Titanic*? The tip of the iceberg or the part hidden below the water line?

Luck Factor

The only difference between an exchange of energy and a loss is Luck Factor 1. Krause (1997) supports this by saying, "If we are lucky however, and experience only a near-miss, then that is just an incident." (p.292)

S.L. Smith (1994) supports Krause's statement by saying that, "Most experts define a near-miss as any incident that could have led to damage to property or injury to employees but, for whatever reason, did not." (p.30) In quoting Bruce Williams, S.L. Smith says that Williams calls near-misses, "warning signs of bigger trouble to come." (p.30)

Smith continues by quoting Fred A. Manuele, CSP, PE, President, Hazards Ltd., Arlington Heights, IL., as saying that "investigations of near-misses should receive higher priority in some cases than actual accidents." (p.33)

Near-Misses: A Reporting System

One of the major weaknesses in numerous safety systems is the fact that near-misses are neither reported nor investigated. Unfortunately, only once an accident has occurred is an investigation conducted and corrective action taken.

As S.L. Smith (1994) says, "But safety educator and author John V. Grimaldi, Ph.D., M.D.R., Calif., argues that the potential for severe exposure, even if nothing happened, is often a better indicator of the effectiveness of a safety program than injury rates." (p.33)

Failure to Report

Most organizations do attempt to get near-misses reported by encouraging employees to report them. After interviewing in excess of 300 employees as to why they don't report near-misses, the reasons given were that when an employee reported a near-miss to his supervisor he was immediately challenged and asked, "So what did you do about it?" On one occasion, an employee reported to me that there was a 12-ft excavation that had been left totally unbarricaded. Another employee had almost driven his vehicle into this ditch. When he reported this incident, his supervisor again challenged him by saying, "So what did you do about it?" In most instances, the employee reporting the near-miss does not have the necessary authority or the resources to take action. In a number of the cases reported by the employees I interviewed, when they did try to take action and things went wrong, they were accused of "operating without authority," or doing something other than prescribed by their task description. Often, doing something about an incident may involve giving instructions or advice to a more senior person and this immediately puts the employee in a difficult position. As a result of this, employees have confided to me that they would rather not report the incidents and not get involved.

Another reason for not reporting incidents is that the employee, if involved in the incident, is often reprimanded for committing the unsafe act or creating the unsafe condition. The employee is blamed and since he had made the effort of reporting how his own behavior could have resulted in an injury or property damage accident, his efforts were rewarded with a rebuff. Fear of discipline and repercussions is the other main barrier to the reporting of near-misses.

The first and most vital ingredient of a near-miss reporting system is that it must be a "no names, no pack drill," or anonymous reporting system. Management should be interested in *what* happened and not *who* messed up. As all these accident ratios indicate, there are always plenty of mess-ups before a serious injury occurs and as long as people are working in an industrial environment, mistakes will be made. The objective of a near-miss reporting system is to find out what actions can be taken to prevent further undesired events before they result in a loss.

Near-Miss Reporting Form

The next step in a near-miss reporting system is to devise a simple, yet effective near-miss reporting form. Most employees and supervisors hate paperwork – especially safety – related forms, questionnaires, etc. As a result of this, a short, brief, near-miss reporting form will reduce the resistance to this "safety paperwork."

The form should contain the following,

- person's name (as an optional)

- the date of the event

- the location of the event

- a brief risk assessment of the event

- a description of what happened

- action taken (if applicable) or recommended

The forms should be readily available throughout the workplace. Ideally, they should be bound in a pocket-size booklet so that the employee always has forms available. If possible, the book should be carbonized so that the employee retains a copy of the incident reported. He can then monitor the follow-ups at his own pace.

Success

The introduction of such a *no blame* incident reporting system facilitated by pocket-size reporting booklets improved near-miss reporting from 10 near-misses per month to approximately 300 per month at one organization. Correct training on how to use the system also boosted its usefulness. Previously, forms were distributed haphazardly and no training was given. A 2-hr training session describing the importance of near-miss reporting, how to use the form, and how to conduct a risk assessment of the incident potential was met with tremendous success and, as indicated, resulted in more than 300 near-misses being reported in the first month.

Flow Process

Every near-miss reporting system should have a flow process showing what happens to the near-miss form that has been handed in for corrective action. The near-miss report should have ranked the level of risk via the mini risk assessment incorporated in the forms, which will determine the level of action that the event will receive. The higher the risk potential and severity is ranked, the higher the priority of rectifying the hazard. The completed actions should be channeled to the safety committees for their scrutiny, and, once all actions are completed, the form is filed for future reference.

Predictive

Accidents that result in injury, property damage, or business interruptions are often preceded by near-misses. The consequences of a contact may often be hidden and, though there is contact, under slightly different circumstances the consequences could have been greater. Identifying and rectifying the causes of near-misses eliminates all probable consequences.

Reporting, recording, investigating, and taking corrective actions on incidents is predictive safety management in top gear. Incidents are warnings of a failure in the system that is about to culminate in a major loss. In many cases, incidents are more important than accidents. The most important words in safety are, "It's not what happened, but what *could have* happened."

Investigating and rectifying causes of high-potential near-misses will eventually lead to a reduction in the number of accidents and their resulting injuries.

As W.E. Tarrants (1980) said,

> It is appropriate to consider possible means for estimating prospective accident experience through some direct measure of *accident propensity*. To be useful for this purpose, it is desirable, but not necessary, to be able to transform measures of propensity directly into estimates of future accident experience, or at least insofar as the objective is to make relative rather than absolute assessment. (p.187)

As is demonstrated by the CECAL Model, the only difference between an accident and an incident is the outcome, which is largely determined by Luck Factor 1.

Risk Assessment

The risk assessment of reported incidents is important. In the past, safety practitioners generally recommended to management that *all near-misses* be investigated. This led to a tremendous workload and a waste of time and effort. Incidents are important, but those with high loss-potential severity and high probability of recurrence should receive priority when it comes to investigation.

A simple risk matrix such as that produced in Model 1.6 (p.21) can be included on the incident reporting form.

The person reporting the incident merely has to check a block as to the loss potential severity of the incident, (or what could have happened), as well as the probability of the event happening in the future.

A risk ranking of the near-misses can then be established. Some incidents can be rectified immediately and the action taken can be entered on the form.

Events that fall in the "gray" area of the matrix, should be investigated. Those that fall in the "black" area (see model 1.6) should receive the same prominence and attention as a fully-fledged accident investigation of an injury, property loss or business disruption.

Example

A 'four by four' that fell from supplies being unloaded almost hit a truck driver. Immediate action taken, "I cautioned the man about standing too close to the action."

S.L. Smith (1994) also references Grimaldi. "Grimaldi counsels safety professionals to rate near-misses in order of their potential for severe exposure and consequence, and tackle the big ones first." (p.34)

Pro-active Safety Controls

S.L. Smith continues: -

> But a growing number of safety experts now agree that near-miss incidents may provide a better measure of the effectiveness of a safety program than tracking injuries and that their reporting and investigation are critical in preventing injuries and fatalities from occurring. Most experts define a near-miss as any incident which could have led to damage to property or injury to employee but for whatever reason did not. (p.33)

Near-miss trends can be calculated and used to initiate pro-active safety program measures.

Results

Model 6.5 shows near-misses reported per month as well as the actions taken. These actions are pre-contact, pro-active safety activities initiated to prevent a recurrence of these warnings, which, under slightly different circumstances, could have resulted in loss.

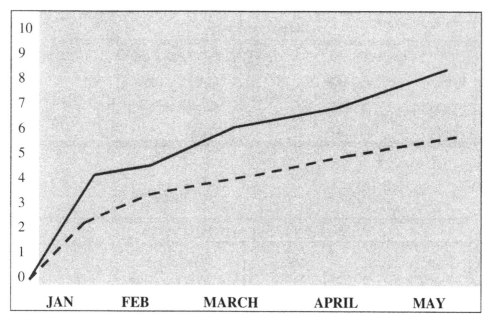

Solid = Near-misses reported.   Dashed = Corrective actions taken.

Model 6.5 – Near-miss reporting and tracking system.

Risk Matrix

Using a near-miss reporting system, a risk matrix analysis was compiled and is reproduced as Model 6.6. This table shows that the majority of the incidents have *medium* loss-severity potential and probability of recurrence was *low*. However, nearly 25% of the incidents had high potential for loss severity and 6% had a high probability of recurrence.

RISK MATRIX					
SEVERITY			FREQUENCY		
LOW	66	**36%**	LOW	105	**57%**
MEDIUM	75	**41%**	MEDIUM	68	**37%**
HIGH	44	**24%**	HIGH	12	**6%**

Model 6.6 – Severity/Frequency Analysis.

POTENTIAL ACCIDENT			
Struck against	6%	Fall-lower	10%
Struck by	30%	Overexertion	1%
Caught in	1%	Electricity	2%
Caught on	4%	Acids	1%
Caught between	6%	Toxic Substances	2%
Slip	20%	Foreign Objects	18%
Fall-same	1%		

Model 6.7 - Risk analysis chart

Potential Accident Type

The analysis also analyzes the potential accidents that could have occurred. It shows that 30% of the accident types would have been *struck by*, 20% would have been *slips*, 18% *foreign objects* and 10% *falls to lower levels*.

Using this information, management can then almost predict the severity, the frequency of recurrence, and the type of the next accident.

This compilation of narrow escapes serves as warnings as to what could happen if it were not for Luck Factor 1.

Ongoing Analysis

The next month's analysis of the same reporting system resulted in 215 near-misses reported with 184 actions being taken.

The analysis of the month of September also showed that 3% of all near-misses had high loss-potential severity and 5% of the incidents had a high probability of recurrence.

The potential accident type changed to potential *struck by* – 28%, followed by potential *striking against* – 19% and then *caught on* – 10%.

Feedback

One of the most important steps in the near-miss reporting system is giving feedback to the employees who report incidents. The most obvious form of feedback is the correcting of the situations that have been reported and the taking of actions to prevent the symptoms from recurring.

Numerous organizations have a monthly lucky draw for near-miss reporting recognition. All the members of the division who have reported five or more near-misses put their names in a hat. A winner is drawn. A small prize is then presented to this person to thank him or her for the contribution. Tremendous enthusiasm can be generated by a simple competition.

As is stated in the *MBO Five Star Safety and Health Management System Introduction Booklet* (1991), "The work of any good manager should be to reduce the 'plenty' of near-miss accidents." (p.7)

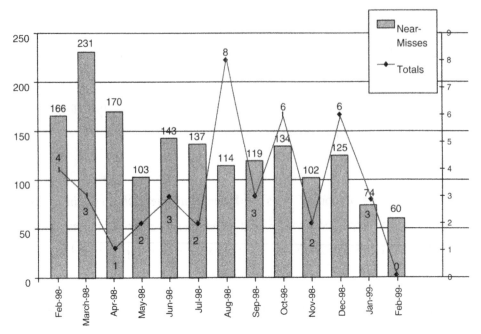

Model 6.8 – Near-misses vis-à-vis injuries

Injuries vis-à-vis Near-Misses

By constantly eroding the base of the iceberg or accident ratio triangle, successes will eventually start to impact the top of the triangle, i.e., the minor and serious injuries.

An interesting example of this was a study carried out in 1998. A near-miss reporting system was encouraged and from a mere 50 to 100 near-misses reported in a month, this figure jumped to 250 in the month of May. In superimposing the reduction of injuries on the number of near-misses reported, it was found that as the near-misses reported increased, so the number of injuries fell. Near-misses reported rose from 75 in January to 166 in February and 231 in March. The injuries reduced from five in January to four in February and three in March. There was a further reduction down to one injury in the month of April. As the number of near-misses reported fell, the injury rate once again continued to climb.

Summary

Near-misses are warnings of a failure in the management system and they should be treated with the same vigor as an accident. High potential incidents should also be investigated, and the basic causes identified and eliminated; this will reduce the number and frequency of accidents.

CHAPTER SEVEN

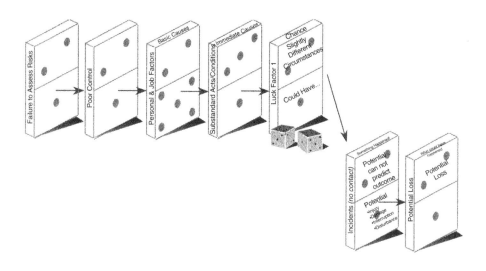

POTENTIAL FOR LOSS

The sequence of events described so far have encountered Luck Factor 1 and have resulted in either a contact with a source of energy or an incident (near-miss).

Traditionally, most near-misses are ignored because "nothing happened," or so it seems. In referring to a number of safety practitioners' accident ratios, it is obvious that there are numerous incidents for every serious injury. It would then seem logical that all these incidents should be treated as seriously as loss-producing accidents.

Misleading

As a young safety advisor, I advised a number of managers and safety professionals to investigate every single near-miss. I was taught to believe that each near-miss could have ended up as an injury. What I was doing was misleading my clients. If I had thought through the process, I would have realized that, due to manpower and cost constraints, it is virtually impossible, as well as totally impracticable, to investigate *all* near-miss incidents.

Potential

In tackling the no-contact events, the keyword is *potential*. What did the event have the capacity to do under slightly different circumstances? The potential is what should determine which of these incidents should be investigated as well as the level of the investigation.

Potential Hazards

Safety practitioners have often debated the expression "potential hazard" and have come to the conclusion that there is no such thing as a *potential* hazard.

They argue that a hazard is a hazard is a hazard. They are correct. The term *potential* hazard really refers to a hazard that has the *potential* to cause harm. Once again, the keyword being *potential* or having the capacity to cause harm.

Loss Potential

As W.E. Tarrants (1963) puts it:

> The existence of an injury or property damage loss is no longer a necessary condition for appraising accident performance. It is now possible to identify and examine accident problems "before the fact" instead of "after the fact" in terms of their injury-producing or property-damaging consequences. This allows the safety professional to concentrate on measurement of loss potential or near-misses and remove the necessity of relying on measurement techniques based on the probabilistic, fortuitous, rare-event, injurious accident. (p.319)

Boylston (1990) refers to incidents as potential problems and also quotes the luck factors:

> Failure by an organization to recognize, evaluate, and implement controls for early warnings of potential problems usually results in a system of reactive approaches. Consequently, there is little if any way to control the magnitude of the problem. Such organizations are "lucky" or "unlucky," depending on the situation. This is no way to manage an organization. (p.103)

Ranking the Potential

The most important aspect of a near-miss is the quantification and ranking of the incident's degree of potential.

The potential could be the potential for:

1. severity of loss

2. recurrence of event

3. number of people affected

These are the common terms of severity, probability, and frequency. A simple method of ranking the potential of a near-miss is to ask the following questions:

- What is the probability of this event occurring?

- If this event occurs, how bad will the consequences be?

- If the event occurs, how often will it be repeated and how many people are exposed?

Safety Solution

It is strongly felt that the solution to safety problems lies within those incidents with high potential, as they are accidents that the organization has not yet experienced. The loss causation sequence has been triggered, but, due to Luck Factor 1, has ended in a warning, a near-miss, or an incident. All management need do, is identify the potential of the incident. If the potential is high in terms of severity and probability, treat the incident as if something *had* happened and institute appropriate control measures.

Crystal Ball

A very simple method of assessing the potential of each near-miss is merely to gaze into an imaginary crystal ball and quote the magic words "it's not *what* happened, it's what *could have* happened." This very simple potential assessment technique will identify those incidents with the greatest potential for loss and which should be treated with the same urgency as loss-producing accidents.

Potential Ranking Sequence

A very simple method of assessing and analyzing the potential of near-misses is as follows:

1. Train and encourage employees to report all near-misses, irrespective of their potential.

2. There should be anonymity of reporting.

3. Assure them that there will be no repercussions and that the system is a "no names, no pack drill" exercise.

4. Issue a simplified incident report form, that includes a risk matrix.

5. Commend employees on submitting near-miss reports.

6. Using a simplified risk matrix system, rank the probability of recurrence and potential severity of each near-miss.

Model 7.1 is such a matrix, rating the probability of recurrence and potential severity from low to high. Those that fall into the gray, or intermediate area, should receive investigation and those that fall into the black area or high-high category should be investigated as thoroughly as accidents that have resulted in a loss. The same accident investigation report should be used and the same diligence applied even though there was no loss. The potential for loss is the guiding factor in this instance.

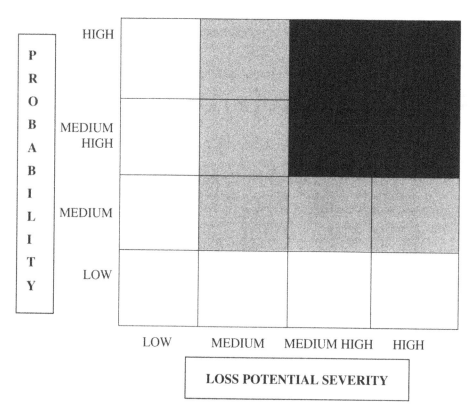

Model 7.1 - An example of a simple risk matrix.

Examples

1. In removing the liners from a mill, an employee left a 3-ft pry-bar lying in the bottom of the mill casing. To remove the liners, the nuts are removed, employees withdraw a safe distance and the mill is rotated, allowing the liners to fall to the bottom of the mill. On this occasion, the liner fell on the pry-bar, sending it flying through the air like an arrow approximately 4 ft from the ground. It hit a supervisor in the lower abdomen. Fortunately, by the time it struck the supervisor, it had lost sufficient energy and it caused only a slight bruise. Although this could be classified as a minor injury, the potential was ranked as extremely high, as was the probability of recurrence.

Witnesses agree that under slightly different circumstances, the pry-bar could have passed right through the supervisor causing serious injury or death.

2.   While hoisting a ramp, a cable broke and the ramp fell, narrowly missing a worker who jumped out of the way. The ramp weighed in excess of 1 ton, and falling from a height of 12 ft, had the potential to cause serious injury to the man. The probability of recurrence was rated high, as the raising of the ramp was a daily activity.

3.   An employee was walking in a walkway where a piece of grating was missing. He nearly fell into a ditch as a result of this hazard. This near-miss was ranked as low potential as the ditch was only 1 ft deep and the probability of recurrence was also ranked low as this was an isolated occurrence of the grid being removed. In ranking the potential of near-misses one could picture the worst-case scenario, which would raise the potential of most near-misses. It is advisable to use the *normal expected loss* approach by asking the question, "What is the probability of recurrence and the loss-potential severity under normal circumstances?"

4.   An employee started work on an electric motor and received a slight electric shock. The potential of this near-miss was ranked as extremely high and the probability of recurrence as medium-high. An electric shock situation has high potential to cause death and this is one instance with high potential that deserves thorough investigation and the implementation of remedial measures. The difference between a slight electric shock and electrocution is once again determined by Luck Factor 1.

5. While lifting a 300-lb. battery, the battery cable broke and the battery fell 4
    ft, just missing a man's feet. Considering the size of the battery and the
    distance falling there was obviously potential for severe injury to the man's
    feet, had he not jumped clear. This near-miss was ranked as medium-high on
    the probability of recurrence as this was a daily activity, and loss potential
    severity was ranked as medium-high. This would rank the potential of the
    near-miss as sufficient to warrant a further investigation.

Summary

Near-misses are warnings that the management system is flawed. Not all near-misses however, need immediate and in-depth reactions as some near-misses have less potential than others do. Assessing the potential of a near-miss gives a clear indication so that investigations can be prioritized. Using a simple risk matrix, each near-miss's potential can be ranked and consequent investigations and follow-up actions prioritized. Near-misses should be treated as friendly warnings. Friendly warnings that have high potential offer a clear indication of what *could* happen under slightly different circumstances, which are normally beyond our control. Management should take heed of these high potential near-misses and institute controls before a loss occurs. Should a contact take place, the consequence of that contact is fortuitous.

CHAPTER EIGHT

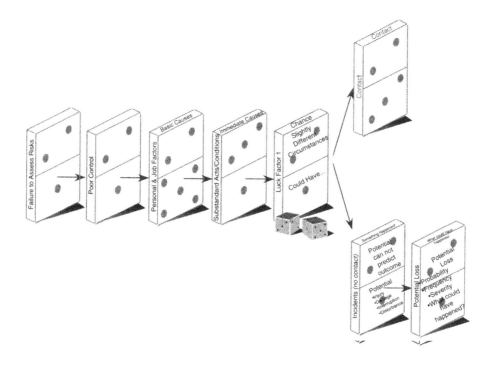

EXCHANGE OF ENERGY

Once an unsafe act or an unsafe condition is imminent, Luck Factor 1 determines whether there will be an exchange of energy. This exchange is in the form of a contact with a substance or source of energy greater than the threshold limit of the body or article. The unsafe act or unsafe condition may result in an incident, where, although there was potential, there was no contact.

The contact, or exchange of energy, is the part of the CECAL sequence that is most closely associated with the loss. The exchange of energy is what injures, damages, pollutes, or interrupts the business process.

Injury

The National Safety Council's (1992) definition of an *injury*, given in *Accident Facts*, further describes this exchange of energy as follows: "An injury is physical harm or damage to the body resulting from an exchange, usually acute, of mechanical, chemical, thermal, or other environmental energy that exceeds the body's tolerance." (p.111)

The accident is the entire undesired event constituting the CECAL sequence. The *contact* part (domino) of the sequence is what causes the actual loss.

Confusion

A number of people refer to the contact phase of the sequence as the "accident." The definition given by Heinrich (1959) is that *"an accident is an unplanned, uncontrolled event in which the action or reaction of an object, substance, person or radiation results in personal injury or the probability thereof."* (p.23)

In explaining his domino theory of accident causation, Heinrich's 4th domino is entitled "the accident" and is explained as "events such as falls of persons, striking of persons by flying objects, etc., are typical accidents that cause injury." (p.23)

To clear any confusion, the *accident* referred to by Heinrich is the exchange of energy or the *contact*, which does the harm.

The National Safety Council (1992) defines an *accident* as *"that occurrence in a sequence of events which usually produces unintended injury, death or property damage."* (p.111) This definition, as well as Heinrich's, leads one to believe that the exchange of energy is the *accident* and not the entire sequence of events. In referring to the Bird and Germain (1992) definition, the *accident* is defined as *the undesired event that results in, a loss to people, etc.* The event is the sequence that leads up to and includes the exchange of energy.

Accident Types

The exchanges of energy are also called accident *types*. A more apt description would be *energy transfers* and Bird and Germain (1992)(from ANSI ZI6.2-1962, rev.1969) quote these as the following:

- Struck against (running or bumping into)
- Struck by (hit by moving object)
- Fall to lower level (either the body falls or the object falls and hits the body)
- Fall on same level (slips and fall, tip over)

- Caught in (pinch and nip points)
- Caught on (snagged, hung)
- Caught between (crushed or amputated)
- Contact with (electricity, heat, cold, radiation, caustics, toxics, and noise)
- Over stress/over exertion/overload (p.26)

The term "types of contact" or energy exchange are far more descriptive than "accident types." In the CECAL sequence, a distinction will be made between the *accident* itself and the specific type of *energy exchange* or *energy transfers*.

Loss Causing

As mentioned, the energy exchange in the sequence of events is what causes the loss. The National Safety Council (1992) describes the *energy transfer* as an *'accident type'* and defines it as, *"accident type describes the occurrence leading to injury or property damage."* (p.111)

In an accident investigation it was found that the contacts that took place were:

1. Over exertion: the weight and movement on top of the utility pole was greater than the compaction of the base of the pole, causing it to topple.

2. Fall to below: the person and the pole fell to the ground.

3. Struck by: the injured person was struck by the pole when it hit the ground and bounced back hitting him.

4. Fall to below: after being hit by the rebounding pole, the worker then fell onto the ground.

The above is a *contact* analysis of the various exchanges of energy that took place during the accident and that caused injury, damage, and business interruption.

Contact Explained

Heinrich et al. (1969), in explaining the Bird/Germain loss causation model describes *contact* as follows:

> The word "contact" appears on the domino at this point in the sequence because an ever increasing number of researchers and safety leaders around the world look at the accident as a "contact" with a source of energy (electrical, chemical, kinetic, thermal, ionizing, radiation, etc.) above the threshold limit of the body or structure, or "contact" with a substance that interferes with normal body process. (p.27)

In further explaining how Bird and Germain intended the contact domino to fit into the sequence, they continued:

> This point in the sequence is also referred to as the contact stage, and applications of the principles of deflection, dilution, reinforcement, surface modification, segregation, barricading, protection, absorption and shielding are examples of counter measures frequently used as loss control tools here. (p.27)

Heinrich et al. (1969) explains Dr. Michael Zabetakis's theory on unplanned transfer or release of energy as follows:

> Most accidents are actually caused by the unplanned or unwanted release of excessive amounts of energy (mechanical, electrical, chemical, thermal, ionizing radiation) or of hazardous materials (such as carbon monoxide, carbon dioxide, hydrogen sulfide, methane, and water). However, with few exceptions, these releases are in turn caused by unsafe acts and unsafe conditions. That is, an unsafe act or an unsafe condition may trigger the release of large amounts of energy or of hazardous material which in turn causes the accident. (p.32) (Dr. Zabetakis's use of the word *accident* here should read "loss.")

Petersen Model

The Petersen Accident-Incident Causation Model, Heinrich et al. (1969) shows the accident or incident immediately preceding the injury or loss. (p.49) The CECAL Model puts the contact immediately before a loss. The Ball Model quoted by Heinrich et al. (1969) confirms that "the concept of energy release is a necessary part of the accident-causation process." (p.53) Ball's theory is that all accidents are caused by hazards and that all hazards involve energy, either due to involvement with destructive energy sources or due to a lack of critical energy needs. Gene Surrey developed a model of accident causation based on the epidemiological model of Suchman. The Surrey Model has three principal stages with two similar cycles linking them. "First, a dangerous situation is built out of a secure situation, and then the danger is released causing injury or damage." (p.54)

Contact Risk Assessment

As most safety writers seem to be in agreement that there is some form of energy release that precedes the injury, damage, or interruption, the consequences of certain types of energy exchange can be predicted by applying a simple risk assessment.

The risk assessment on the contact would be the answer to the question, "If it happens, how bad would it be?" To estimate the frequency of the energy exchange ask, "What is the probability of recurrence of this exchange of energy?" Ranked on a scale of low, medium, and high, a high-high rating would indicate greatest potential loss from that particular contact.

Agency

The *agency* is the piece of equipment or object closest associated with the loss and is defined by Simonds and Grimaldi (1963) as *"the substance, object, radiation, or person most closely associated with the accident's occurrence."* (p.178)

They list a few general groupings of agencies, which include:

- animals

- boilers and pressure vessels

- chemicals

- conveyors

- dusts

- electrical apparatus

- elevators

- hand tools

- highly flammable and hot substances

- hoisting apparatus

- machines

- mechanical power transmission equipment

- prime movers and pumps

- radiation and radiating substances

- working surfaces and miscellaneous (p.197)

The National Safety Council (1992) defines the *agency* or *agent* as *"the principal objects such as tools, machine or equipment involved in the accident and is usually the object inflicting injury or property damage."* (p.111)

The *agency* is therefore a key factor in the exchange of energy. It is the agent that is responsible for the energy transfer to the recipient who consequently suffers a loss in the form of injury. The loss could be damage or interruption.

Agency Part

The *agency part* is that part or area of an agency that inflicted the actual injury or damage. For example; a worker was ripping planks on a 30-cm circular saw. To speed up production he removed the machine guard, thus exposing the blade. During the cutting process, he was distracted and the blade cut his finger.

In this case, the agency is the circular saw and the blade would be classified as the agency part.

Agency Trends

A trend analysis can be made by listing the agency that causes the injury, disease or damage. Trends can be used to establish which agency is responsible for the majority of losses as a result of contact.

Two Types of Agencies

There are 2 major classifications of agencies that are involved in the exchange of energy.

1. *Occupational Hygiene Agencies:* Occupational hygiene agencies are those items that are closest to and cause the illness or the occupational disease. They include:

- gas

- heat

- noise

- fumes

- radiation

- ergonomics

- lighting

- chemicals

- etc.

2. *General Agencies:* These include the following:

- walkways

- machines

- ladders

- sharp edges

- power tools

- machinery

- equipment

Occupational Disease

An *occupational disease* is defined by McKinnon (1995) as:

> An occupational disease is a disease caused by environmental factors, the exposure to which is peculiar to a particular process, trade or occupation and to which an employee is not normally subjected or exposed to outside of or away from his normal place of employment. (p.27)

Fatal Statistics

In the National Safety Council's *Accidents Facts*, 1993 edition, the leading causes of accidental death for the year 1989 are ranked by the accident types and are listed as follows:

- motor vehicle accidents

- falls

- poisoning by solids and liquids

- fire and burns

- drowning, suffocation by ingestion (p.9)

In listing the number of deaths due to injury they also classify a number of accident types or manner of injury as well as agencies.

In their accident type "poisoning by solids and liquids," they list certain agencies such as analgesics, barbiturates, tranquilizers, antibiotics, anti-infectives, alcohol, cleansing and polishing agents, etc. (p.10)

CONTACTS (Energy)		LOSS and COSTS (Estimated Loss)	
1) Driver struck against cab of #5 Battery Motor.		Injury to employee. Property damage to # 5 Tow Motor.	$28,000 10,890
2) #5 Tow Motor struck against other carriages.		Property damage to motor and carriages.	3,000
3) #5 Tow Motor and carriage fell to a lower level.		5 Hours of production lost.	

Model 8.1 - Showing a contact analysis of an accident investigation.

Contact Control

Management has three opportunities to exercise the safety management function of *control* within the cause, effect, and control of accidental loss sequence (CECAL). The first opportunity is *pre-contact* control, which is setting up systems, identified by risk assessment, to manage the safety activities on a day-to-day basis. This reduces the basic causes of accidents and in turn leads to a reduction in unsafe acts and unsafe conditions. This pre-contact effort will eliminate the accidental exchange of energy that causes the loss.

The second opportunity for control is during the *contact* stage. Control during the contact stage of the accident sequence can only be directed at minimizing or averting the amount and type of energy exchange. Contact control does not stop the sequence of events that leads up to the exchange of energy. It only deflects or transfers the amount of energy in another direction so that it does not cause harm to the body or structure. As Bird and Germain (1992) describe contact control:

> Control measures that alter or absorb the energy can be taken to minimize the harm or damage at the time and point of contact. Personal protective equipment and protective barriers are common examples. A hard hat, for instance, does not prevent contact by a falling object, but it

could absorb and/or deflect some of the energy and prevent or minimize injury. (p.26)

Many safety programs are focused entirely around the wearing of personal protective equipment, which, as indicated by Bird and Germain, will deflect an amount of energy but will not prevent the undesired event from happening. Nor will it stop the agency being in a position to transfer that energy to a person, piece of equipment, or the environment. Contact control is normally resorted to as an absolute last resort to minimizing the degree of energy exchange.

Example

Underground mining creates numerous hazards. A constant threat to underground miners is the possibility of head injuries caused by falling rocks and stones from the roof of the various tunnels in which they work. Although every precaution is taken to secure the roof, rocks and pebbles do fall and as a secondary measure, the wearing of hard hats is compulsory underground. This is a form of contact control.

Contact control should never be relied on as the first line of defense. As explained earlier, losses may be as a result of multiple contacts within the loss causation sequence. Thorough and effective accident investigation should identify all exchanges of energy that took place as well as all agencies and agency parts that were involved in causing the loss.

Contact Consequences

Contact with a substance or source of energy can cause numerous forms of loss. Most common are:

1. injury to people

2. damage to property and equipment

3. business interruption

4. damage to the environment

Once an exchange of energy takes place, the end result and extent of loss is difficult to predict. As with the outcome of an unsafe act and/or unsafe condition, the outcome is largely fortuitous. By the time the contact takes place, the sequence of events (like the fall of the dominoes) is already progressing so fast that intervention at this stage is too late and some form of loss is imminent.

Summary

This unplanned and unpredictable exchange of energy in the form of contact causes losses. There are numerous types of contacts and the degree of consequence cannot be predicted. The outcome of this energy exchange could be minor, major or catastrophic. Luck Factor 2 determines the outcome of this contact.

CHAPTER NINE

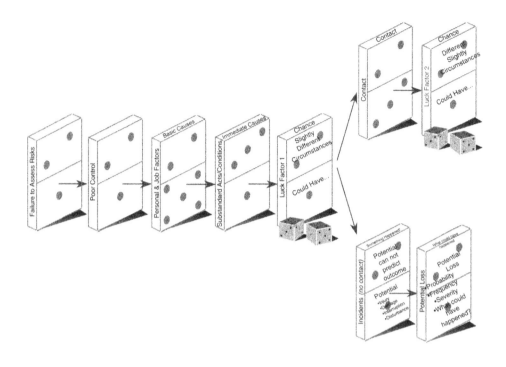

LUCK FACTOR 2

Introduction

The failure to assess risks and institute the necessary controls causes basic causes to exist. They in turn create the unsafe act and unsafe condition which, as Luck Factor 1 would have it, results either in an incident (no loss) or a contact (loss).

The contact is what causes the loss and the outcome of an exchange of energy is determined by Luck Factor 2.

Types of Loss

Once the exchange of energy takes place, the outcome cannot be accurately predicted.

The result of this exchange of energy, if above the threshold limit of the body or the substance, could be:

- personal injury

- property damage

- business interruption

- or a combination of two or all of the above

There are numerous other sub-categories of loss such as environmental pollution, business disruption, delays, etc., that are all some forms of accidental loss. Three main categories are used here.

Tradition

Traditionally, most safety efforts were directed at one specific type of loss – injury to people. Only with the introduction of loss control and total loss control, techniques were efforts focused at other consequences of an undesired exchange of energy.

Elsabie Smit (1996) et al., quotes McKinnon, "After the contact, luck again plays a role in determining the degree of severity of the contact. The outcome could be injury to people, damage to property or process interruption." (p.281)

Example

As the National Occupational Safety Association (NOSA) (1992) puts it:

> Why is it that some people write off their motor vehicles in an accident and get out of the wreck without a scratch or a bruise and on the other hand somebody has a similar minor accident (for instance bumping into a lamp post), minimal damage is caused to the car, yet the person dies because he broke his neck on impact? (The safety belt factor is excluded from these examples). (p.9)

Accident Ratios

In referring to the Bird and Germain accident ratio discussed previously, the determining factor among the 30 property damage accidents and the 11 that resulted in personal injury is largely fortuitous and is as a result of the domino entitled Luck Factor 2.

SWiFT

SWiFT is a hazard identification method used to endeavor to predict the outcome of an undesired event. It is a systematic system of hazard analysis that uses brainstorming techniques to consider the deviation, the hazard, and the cause, and endeavors to determine the consequence, if it should happen. This predictive method fits in well with risk assessment applications and reduces the uncertainty of outcomes of undesired events.

Examples

Imagine a person walking below an unguarded scaffold when a brick is accidentally bumped and falls to the ground. In the first instance, the brick falls to the ground and causes neither damage nor injury but minor process interruption, as the brick has to be picked up and returned to its original position.

In the second case, the same undesired event occurs and the brick falls, this time breaking and creating a loss in the form of damage to material.

In case three, the exact same undesired event takes place, the brick falls from the scaffold and this time hits a worker who happens to be passing by below. The exchange of energy causes minor injury.

In all the three examples given above there was an exchange of energy, yet the three outcomes or losses were totally different. This is as a result of Luck Factor 2.

It is extremely difficult, and in some cases impossible, to determine the outcome of an undesired exchange of energy. One factor that is inevitable is that the exchange of energy *will* result in some form of loss, the detail of which is largely chance.

More Examples

The following four examples are real-life accident case studies. The contact type was "contact with electrical current." The actual contact is described exactly as it was reported as well as the consequences. In all four scenarios, one would reasonably expect there to have been an electrocution but, to emphasize the unpredictability of the outcome of accidents, the scenarios are given to show how Luck Factor 2 determines the result.

Case Study 1: "I was climbing the ladder into the mezzanine and felt an electric shock."

The medical diagnosis was a mild electric shock and the patient was in no distress at the time of examination. He experienced no numbness or tingling. The vital signs were stable and within normal limits.

Case Study 2: "We were stripping steel scaffolding and there was a 440-volt cable on the ground. I placed a walk-board on the ground to kneel down and grabbed the scaffolding with my right hand. I brushed the cable with my left hand, and got shocked."

The result of this contact with a source of electricity was diagnosed as possible mild electrical shock. No chest pain, no nausea. Vital signs normal.

Case Study 3: "I was switching cars when the car hit an electrical cable and shocked me."

The result of this contact was mild electrical shock to the person's body with no apparent burns to the body.

Case Study 4: While cleaning contacts in a 480-volt main breaker cubicle, an electric arc occurred causing a first-degree thermal burn to the electrician's right wrist and a minor flash burn to his neck below his chin.

The electrician was treated for minor burns to his arms and neck and was released to return to his normal duties and completed the scheduled shift.

In all four accidents, the parties were lucky not to have been electrocuted, as even 20 milli-amperes can cause electrocution. Although the exchange of energy was the same, Case Study 4 produced different results in the form of minor injuries. It can be agreed that all scenarios had high potential for loss and that unsafe conditions or unsafe acts had resulted in contact with the source of electrical current but the outcomes were relatively minor due to Luck Factor 2.

Summary

Once an inadvertent and unplanned exchange of energy takes place, the outcome is unpredictable and could result in either injury to persons, damage to property and equipment, or some form of business disruption. Pollution and other forms of loss may also occur as a result of this energy transfer.

The results of a contact are unpredictable and the outcomes are normally as a result of chance or Luck Factor 2.

CHAPTER TEN

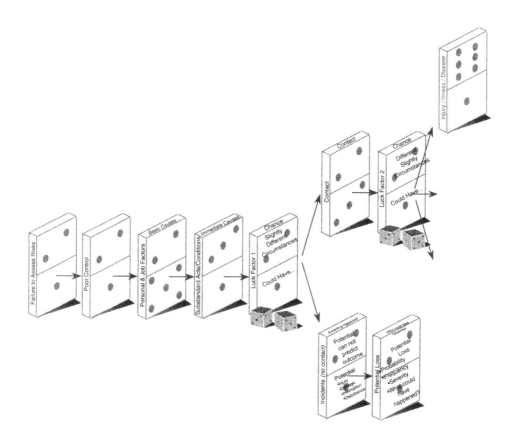

INJURY, ILLNESS, AND DISEASE

The contact that leads to an exchange of energy can result in business interruption, damage to machinery, property, materials, and vehicles, or lead to personal loss in the form of injury or occupational disease to an employee.

Luck Factor 2 determines the outcome of the exchange of energy and this chapter will discuss the injury, illness, and disease outcome as a result of contact with a source of energy.

Main focus

The injury has traditionally been the main focus in loss-producing events. Most safety practitioners still focus on the injury to people as their main concern. The perception of safety is normally the number and degree of injuries that occur in a workplace.

Traditionally, the injury was the only tangible proof that something had gone wrong with the process. It was also tangible proof of the consequences of either an unsafe work environment or an unsafe behavior. Injuries are more spectacular. They are instant, graphic results of the consequences of an accident.

Page Me!

Even today, numerous safety personnel insist on being paged the minute a person is injured and substantial medical response is needed. These same safety professionals are not paged when a significant property-damage accident occurs or when there is a major business interruption due to a loss-producing event. The injury seems to be an emotional aspect of safety that demands attention. I have often asked the question, "What help can a safety person be *after* the event that caused the injury?" Or is the safety practitioner more interested in determining whether the injury is severe enough to be recorded as either a reportable, under legal requirements, or a lost-time injury that will significantly affect the "safety record" and reflect poorly on his or her safety performance?

Costs

In numerous instances, the injury produced by an undesired event is not the most expensive consequence. Property damage, loss of production, business interruptions, etc., all caused by a similar sequence of events, normally cost more than injuries themselves.

Post Contact

In the Cause, Effect, and Control of Accidental Loss (CECAL) sequence, the *injury* domino is in the *post* contact or reactive stage. Reacting to injuries is a traditional safety approach. Some organizations have a standing rule that as soon as an injury occurs, a procedure must be drawn up to cover whatever the injured employee was doing at the time of the accident. This may not be necessary, unless the lack of a procedure caused the accident. Although necessary, *post-contact* activities are far less efficient than *pre-contact* activities, such as assessing the risk, instituting the necessary controls, and ensuring by an ongoing measurement and evaluation system that the controls are in place and are working.

International Misconception

The misconception that exists internationally is that a high number of injuries indicate "poor safety" and that an absence of injuries indicates "good safety." Quotations from authorities in this field will be given to show that injuries are used internationally to judge safety-management success. They are used for comparative purposes between organizations within the same state, between states, and even between organizations in different countries.

Having already analyzed the CECAL sequence up to this domino (injury, illness, and disease), it is obvious that the injury is a consequence that has already been determined by two luck factors. They determined the outcome of the action or condition as well as the outcome of the exchange of energy. The injury therefore, is largely fortuitous and should not deserve the amount of attention that it gets in the safety management process. Admittedly, the injury is physical bodily harm to a friend, colleague, or fellow worker and one must be sympathetic to the pain and suffering that this person will undergo as a result of the injury. The emotion and feeling for the injured person should not cloud our vision and understanding that the injury was merely one of the events in a chain reaction that could have been prevented by proper controls.

Victim

The injured person is a victim in the safety game. Being injured as a result of an accident does not necessarily mean that the injured person was the cause of the accident. In numerous organizations, the victim is automatically blamed for being injured. This means that, irrespective of the true and basic causes of the accident, the victim bears the brunt of the accusation.

Cleverly worded accident investigation reports can point fingers at the victim irrespective of the circumstances of the accident. The fact that the victim may have committed an unsafe act still does not necessarily justify blaming him or her for the accident.

Some of the classic excuses that I have read on accident investigation reports are so simplistic and obviously blame finding that it is amazing management has accepted them.

For example: a construction worker was working on the ground level where his task was to load bricks onto a hoist that then hoisted these up to the bricklayers on the next level. One of the bricklayers on the first level had not erected a toe-board on the scaffolding. During the course of the day he accidentally kicked a brick, which slid off the scaffold and fell onto the worker below, severely injuring his shoulder. On reading the accident investigation report, I picked up that the builder foreman had given cause of accident as "employee took up an unsafe position and a brick fell on his shoulder causing injury." Although it may sound ridiculous, this is an example of cop-outs or finger pointing directly at the victim. Another example is taken from a deep-level gold mine where an employee was walking along a drift and a piece of the concrete lining the back wall fell on his head, causing him to fall and strike his head on the ground. The accident report read as follows: "The employee failed to expect the unexpected." The injured employee is not necessarily the person who caused the accident. Thorough accident investigation will find the true causes of the circumstances that led to the injury.

Occupational Diseases

Occupational diseases normally manifest over a longer period than occupational injuries. They are less spectacular and normally less apparent than an injury and have, regretfully, not received the importance that they should have in the safety-management system. Perhaps a reason for this is that the words "accident" and "injury" have been made synonymous by safety professionals in the past. This has led the general public to believe that *safety, accidents,* and *injuries* are all together, as one, and occupational diseases and illnesses have been almost disregarded.

Injuries are normally tangible consequences of an undesired event and happen quickly. Occupational injuries and diseases are long term and not so obvious. Noise-induced hearing loss, for example, normally occurs over a long exposure period and is not discernible by either the employee or his work colleagues. By the time the employee has suffered a hearing threshold shift, his hearing will have been permanently impaired to a degree that will affect his work and social life. It is easier to convince employees to wear fall protection when working at heights, as they can perceive the immediate consequences of falling and striking the ground. The consequences are usually short term and immediate. Convincing workers to wear respirators eight hours a day to prevent them inhaling dust, which contains silica, asbestos, or other harmful agents, is far more difficult. Since there are no *immediate* consequences, this leads to a false sense of security, which often prevails in safety, that is, "Nothing will happen to me."

Synonymous

The words "accident" and "injury" have been incorrectly used in the past. Accident and injury have become synonymous and therefore the cause, effect, and control of accidental losses have also become confused. The injury is totally different from the contact with a source of energy as are the other events leading to the contact. Heinrich et al. (1959) explains the confusion as follows:

> To the early safety practitioner, the terms "accident" and "traumatic injury" were almost synonymous. While occupational diseases, fire, and property damages were philosophically associated with industrial safety, actual accident prevention practices through the years have largely been devoid of these considerations and are quite injury-oriented. Thus, the word "injury" has been most frequently used to mean bodily damage or harm through traumatic accident. (p.28)

One of the biggest changes that the safety profession could make is to use the term *injury* in its correct context. An *injury* is not an *accident;* it may result from an accident.

Injury Defined

Heinrich et al. (1959) defines *injury* as follows*:* *"Injury as used in this factor of the sequence, includes all personal physical harm, including both traumatic injury and diseases, as well as adverse mental, neurological, or systemic effects resulting from workplace exposure."* (p.28)

In *Accident Facts*, the National Safety Council define*s injury* as: *"Injury is physical harm or damage to the body resulting from an exchange, usually acute, of mechanical, chemical, thermal, or other environmental energy that exceeds the body's tolerance."* (p.111)

Mark A. Rothstein (1998) gives the Occupational Safety and Health Administration (OSHA) definition of an *occupational injury* as: *"Any injury such as a cut, fracture, sprain, amputation, etc., which results from a work accident or from exposure in the work environment."* (p.183)

OSHA further defines an *occupational illness* as*:* *"Any abnormal condition or disorder other than one resulting from an occupational injury, caused by exposure to environmental factors associated with his employment."* (p.237)

The above definitions indicate that injury is some form of harm caused to a person's body. The harm could be acute or chronic and is normally caused by an accident. It is clearly a separate entity from the accident, which is the *event* that caused the harm.

Statistics

In safety, most statistics concern injuries. Statistics based on injuries are also a traditional approach to justifying the need for better safety measures. It is pleasing to note that recently statistics include damage to equipment, loss of production and cost associated with environmental losses. These are also caused by accidents.

Injury statistics are used as attention-getters. They quantify the number and extent of the injuries and diseases caused by workplace accidents and accidental exposures.

Accident Facts

Perhaps the most useful injury statistics are in the booklet *Accident Facts* (1993), published by the National Safety Council. Once again the word "accident" has been used to mean "injury" as this book contains statistics concerning injuries including fatal injuries as a result of accidents. On page 4 of that edition, the explanation is that "the term 'accidents' covers most deaths from injury and poisoning," which explains that the information is really concerning accidental *injuries.*

Also on page 4, the National Safety Council explains that in 1992 there were 83,000 deaths as a result of all accidents. These figures exclude homicides, deaths, and war deaths.

Motor vehicle accidents, falls, poisoning by solids and liquids, fires and burns were the leading causes of accidental deaths in the United States in 1989. (p.9) They report 3,300,000 disabling injuries, which are attributed to all industries for the year 1992. (p.34) Concerning work deaths, the publication reports 8,500 deaths as a result of work accidents during the year 1992. Total cost of injuries as a result of accidents is given as $115.9 billion for the year 1992. Each disabling injury is costed at US$27,000. Sixty five million man-days are lost as a result of these accidents. (p.35)

"While You Speak"

On page 25 of the 1993 edition of *Accidents Facts* an attention-getting quotation called "while you speak" is given:

> While you make a 10-min safety speech – 2 persons will be killed and about 330 will suffer a disabling injury. Costs will amount to US$7,600,000. On the average, there are 9 accidental deaths and about 1,950 disabling injuries every hour during the year. (p.25)

The Gold Barons

In an Internet article by Darren Schuetter, published on November 11, 1998, he refers to mine fatalities and injuries in South African gold mines.

> In the first 93 years of this century, about 69,000 miners died in accidents and more than a million were seriously injured. The 1986 Kinross mining disaster, when 177 workers were killed as a result of a polyurethane fire, was a notorious example of the color bar that plagued the industry. (p.2)

The report continued by saying:

> The Commission conceded that mining is an inherently dangerous job, but it found that too often "profitability ranked higher than peoples' lives" – as evidenced by the asbestos scandal and the continued use of polyurethane in mines long after the dangers had become known. (p.2)

World War II

W. Tarrants (1980) also gives attention-getting statistics concerning people killed in action.

> During World War II about 375,000 persons were killed in the United States by accidents and about 408,000 were killed by war action. From these figures it has been argued that it was not much more dangerous to be overseas in the Armed Forces than to be at home. (p.281)

Accident Ratios

Although the accident ratios have been discussed previously, it is opportune to revisit them, as injuries form a major component of the ratio. It will be remembered that Heinrich was the first to compile and publish the accident ratio. To refresh, he estimated that for every serious injury there were 29 minor injuries. The Health and Safety Executive (Great Britain) compiled its ratio in 1993 and found 11 minor injuries for every serious injury. Frank Bird and George Germain propose the Frank E. Bird Jr. ratio that indicates for every serious injury there were 10 minor injuries.

As mentioned previously, it is always the serious injury that gets management's attention. The minor injuries are normally referred to as first aid cases or injuries not requiring medical treatment and often these injuries do not play a significant role in determining safety program effectiveness or non-effectiveness.

The accident ratios therefore indicate that approximately only 2% of all accidents result in injury. In the Bird ratio, 15% of all undesired events result in serious injury and only 1.7% in minor injury. This shows that injuries are but a small percentage of total losses caused by accidents.

Model 10.1 - Shows the accident ratio for 10 months of 1996 (left) and the accident ratio for the month of October 1996 (right).

Reportable

Legislation normally defines what types of injury are reportable and what types need only be recorded in some form of register or log. These definitions differ among legal agencies, Workers Compensation, and insurance carriers. They differ from country to country and are therefore not a reliable form of measurement at all. The very criteria that determine whether they are reportable are inconsistent.

As these injuries and diseases are also the result of two luck factors, they are not a good indication of the work conditions or work practices at a workplace at all. Legal agencies would be far better off measuring the degree of controls that the organizations have put in place rather than basing the assumption of safety efforts on the degree of injury.

Injury Types

There are a number of injury types that are usually grouped into 10 or 12 major categories. These give an indication of the type of injury being experienced for statistical analysis and tracking purposes.

Some of the main injury types are as follows:

- traumatic amputation

- asphyxia

- burn (heat or chemical)

- contusion (bruise)

- wound (laceration/abrasion)

- skin irritation

- dislocation

- electrical shock

- fracture (open/close)

- frost bite

- heat exhaustion/stroke

- hernia

- inflammation

- sprain/strain

- multiple injuries

- puncture

- foreign object

Although these classifications are by no means a complete description of the nature of injury that could be experienced, they are useful in determining what type of injury is most frequently caused by accidents. Some form of trend could be plotted using the nature of the injury as a category. The nature of injury would vary from industry to industry and certain injury types are more frequently experienced in different industries.

Measurement

Injuries are still used as the measurement of safety or the lack thereof. Organizations that experience fewer injuries than others are lured into a false sense of security by assuming that they are safer. Good safety controls are often *assumed* when an organization has low injury rates. Injuries are a poor indication of safety performance and are an even poorer indication of safety success. Injuries indicate a failure in the system and are more likely a measure of failure than of success.

The tradition has always been to rank a company's safety performance as well as its management's safety performance by the number of injuries experienced over a specific period. Internationally, this precedent has been set. It would prove to be exceedingly difficult to convince managers that the injuries they are using to measure their safety management controls are, in actual fact, merely indications of either good or poor luck. This is discussed in the Luck Factor 1 and Luck Factor 2 chapters.

Leading organizations and international companies quote injury statistics in their annual report to their boards of directors. They describe the disabling-injury incidence rate and the lost-time injury frequency rate as "the internationally accepted measure of safety performance." Although it is good that the boards of directors are receiving an annual safety report along with the financial report; it is time for more accurate measurement of safety controls to be used.

Legal Requirement

The reporting of certain injuries arising out of and during the course of normal employment is also a legal requirement in most countries. Most organizations require the reporting and recording of injuries and these are then tallied and used to compare one industry with the industrial average or with other industries and companies belonging to the same organization.

V.L. Grose (1987) puts it in a nutshell when he says:

> Lots of thought and effort have been applied to managing risks over the past 30 years, but it has been only partially effective. Classical risk management's main drawback thus far has been its inability to participate as a full member of the top executive's team because it has not progressed much beyond the moral argument, "no one should suffer losses."
>
> To achieve their rightful status with top management alongside benefit managers, risk managers must merge the "flesh" or specialized activities of traditional risk management, with the "skeletons," or all-encompassing framework of the systems approach. The result can be called Systems Risk Management. (p.11)

Competitions

Traditional safety competitions involve the comparison of injury rates between departments or companies or work sections. Prizes are then awarded to those divisions that have been injury-free or that have had fewer injuries. These incentives, which are still in practice today, have a major drawback especially if they are focused entirely on number of injuries. As Geller (1996) is quoted: "Often incentives for fewer injuries, for instance, can reduce the reported numbers while not improving safety. Pressure to reduce outcomes without changing the process (or ongoing behaviors) often causes employees to cover up their injuries." (p.27)

Mark A. Friend (1997), writing in *Professional Safety* continues the demise of traditional safety incentives:

> A typical safety incentive program is a prime example of how a firm creates an environment that fosters incidents (accidents) under such a program, employees (or work teams) who do not report any incidents (injuries) for a specified period of time are eligible for some reward. In this scenario, employees are encouraged to *not* report what is actually occurring. (p.36)

He then continues and asks whether typical safety incentive programs do harm or good. He summarizes the answer by saying that "typical programs rewarding employees for hiding facts – by encouraging them to not report incidents [injuries]." (p.36) (author's note – the words in brackets have been added so that the reader can distinguish between the concepts of *accident, incident* and *injury* as defined in earlier chapters).

When employees of an organization were interviewed, they admitted that there was a monthly safety bonus for being "injury-free." Paying safety bonuses in the form of cash incentives also leads to injuries being hidden and suppressed wherever possible.

Example: Questionnaire – Twenty-three subjects responded to a 15-question questionnaire.[1] Two of the questions were directly pertinent to injury reporting. Question no. 1 asked, "If you work for a company that had worked 16,000,000 injury-free hours and you were injured in an accident, would you report it?"

Ten respondents said they would, three said 'no' they wouldn't and the other 10 said that, depending on the severity, they would either report it or not.

Question no. 10 asked, "If you were paid a monthly bonus of $150 cash to be free from injuries, would you report a personal injury?"

Three of the respondents replied 'yes,' seven replied 'no' they would not report it and the majority, 13 of the respondents, said that, depending on the severity, they would not report it.

In the chapter entitled "How we traditionally have analyzed" Dan Petersen (1996) also indicates the weakness of using the number of injuries as a measurement of safety:

> At what point do accidents become a valid measure to judge performance? Actuaries state that only when about 1,129 accidents have occurred would they begin to judge the unit that generated those 1,129 to be so believable that future rates could be based on those figures. (p.15)

At the end of the first paragraph he states that, "Using a frequency rate is a touch less ridiculous than this (a monthly fatality rate)" and asks the question, "So what do we use to measure the safety performance of line managers?" He answers the question by saying "Almost anything but accident statistics." He continues, "How about an activity measure: did the manager do those defined activities that he or she agreed to? The activity measure would at least ensure some validity in the measurement."(*Analyzing Safety Systems' Effectiveness*)(p.16)

Writing in *Techniques of Safety Management*, Dan Petersen (1978) quoted further inaccuracies in injury reporting:

> The Department found significant problems and inaccuracies. For example, audit tests of 105 mines indicated that there could have been 118 more disabling injuries in addition to the 283 reported by the mines – an error rate of over 63%. It was also found that there could have been 108 more non-disabling injuries than the 139 reported by these companies – an error rate of 77%. (p.129).

Dan Petersen and James Tye had obviously had discussions on under reporting and using injuries as a measurement of safety when Tye replied to an article in *Professional Safety* in June 1995 by saying "since the under reporting of accidents in the UK runs at about 30% (and 80% in some U.S. industries), frequency rates, frankly, are irrelevant." (p.12)

In Chapter 13 of *The Measurement of Safety Performance*, W.Tarrants (1980) discusses major issues and identified safety problem areas. Concerning the use of injuries as a measure he states that: "Measures are needed that predict, not simply record, accident recurrences. Historical records of accidents mean nothing unless they can be used for prediction and control purposes." (p.235)

He concludes by saying:

> A measurement problem exists in that low injury frequency rates tell nothing about the potential for catastrophe. For example, one plant experienced a $45 million accident loss, with its walls covered with safety awards received based on a low frequency rate. Better measurement techniques are needed to identify loss potential. (p.237)

Most measurements using injuries are also based on the degree of severity of the injury and rely on the honesty of the reporting system. Most incentives drive injuries underground and management is lulled into a false sense of security thinking that its safety effort is having the desired effect.

In one organization where the semi-annual production bonuses were dependent largely on the number of injuries experienced, an employee worked for 6 hrs on a twisted ankle before finally collapsing in agony. When interviewed, he said he had tried to conceal the injury so that his team would not lose their portion of the bonus. His boot had to be cut off his swollen ankle before he could be treated. Peer pressure, team pressure, and fear of letting the team down and having them lose their bonus drove this man to walk on his ankle for 6 hrs before collapsing.

Thomas A. Smith (1997), in his article "What's wrong with safety incentives?" makes the statement that:

> Safety incentives/rewards create competition and fear within an organization. Although many believe that competition is positive, research dispels this myth. External competition will always exist and it does drive business to perform better. Internal competition, however, is not positive. It creates winners at all costs. This is true in both production and safety. For example, an employee may not report an injury because he/she feels they are spoiling the department's record, which would mean no reward. (p.44)

Injury Rates

The number of injuries is normally compiled and expressed as a percentage, or number of injuries per million man-hours. The disabling-injury incidence rate is the percentage of employees suffering disabling injury per year. The lost-time injury frequency rate is the number of employees suffering lost-time injuries over a one million man-hour period.

More sophisticated injury statistics can be compiled by using the disabling-injury severity rate, which records the number of shifts lost per million or per 200,000 man-hours and by combining the severity and injury rate one can produce a lost-time injury index.

It has taken the safety professionals many years to teach management that these statistics are the method to measure safety performance. Very few people who are exposed to these statistics actually know what they mean. Dr. Mike Manning (1998) explains how to work out the incident (the correct term is incidence) rate:

The incident rate is figured by multiplying the yearly total number of injuries and/or illnesses by 200,000. The number 200,000 is an arbitrary number used by the safety industry to represent 100 full time workers working 40 hrs per week, divided by total workers' hours for 50 weeks a year. It is a simple measuring device that any industry can use to compare itself against its peers. (p.32)

One of the shortcomings of using injuries as comparisons is that certain industries and mines have more hazardous conditions and it is therefore unfair to use the number of injuries as a comparative measurement.

Injury Analysis

The number, type, and body part injured are post contact statistics that can be analyzed to plot certain trends.

An example of the executive summary of an injury analysis report that covered 12 months of injury experience is as follows:

1. The first analysis was to identify employees who had been injured more than once during the last 12 months.

2. The next analysis was to confirm the department's DiiR, and through this process, to determine the DiiR of individual departments and teams.

3. The day of the week and the number of injuries were analyzed and are shown in a pie chart format.

4. The time of day at which injuries occurred was analyzed and the results shown on a bar graph for the 23-hr cycle.

5. Injuries per department within the organization were also analyzed and displayed as total injuries per department on a bar graph.

6. Total injuries as per job classification were analyzed and shown on a bar graph. The accident (contact) types were analyzed in relation to the frequency and severity of injuries, namely disabling and non-disabling injuries.

7. The type of injury and the body part involved in these injuries is shown on two different graphs.

8. The total organization's injuries analyzed per first and second half of the year is shown.

9. Injury classification that constitute the total injuries.

Report Summary

The report concluded with the following summary:

1. Total number injuries reported was 81.

2. Fifty three percent of the department's workforce reported an injury during this period.

3. Seventeen employees were responsible for reporting 45 injuries.

4. Days of week injuries being reported showed nothing significant.

5. Most injuries occurred on day shift between 07:00 a.m. and 12:00 noon.

6. Laborers reported 52% of all injuries.

7. The leading accident type was over exertion and the most severe was electric shock.

8. Parts of the body most frequently injured were back, hand and fingers, arms, and eyes.

9. Sixty nine percent of injuries were recordable and 31% were first aid injuries.

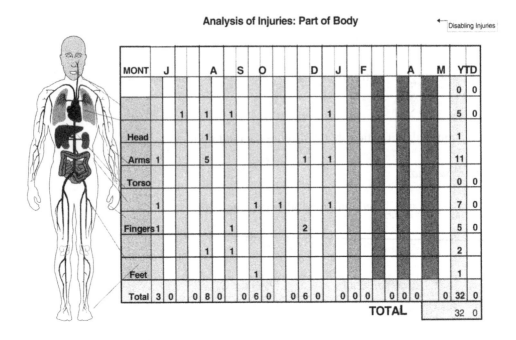

Model 10.2 - Shows the number of injuries occurred to the body parts during the 12-month period. The model clearly shows that arms were injured more often than other parts of the body followed by back, hands and fingers.

Conclusion

Injuries receive a lot of prominence in the safety-management process. They are tangible consequences of the effects of accidental loss. Only a small percentage of accidents end up in injuries. Accident ratio studies have indicated that there are often more minor injuries experienced for every serious injury and that for the same serious injury there have been numerous property-damage accidents and a large number of near-misses.

Regretfully, injuries are still being used to measure safety successes and management's safety leadership. Due to the fact that they are largely fortuitous, as shown by the CECAL sequence, plus the fact that there is doubt as to the accuracy of the reporting of injuries, they are a poor indication of safety. Numerous safety-incentive schemes are a farce as they merely encourage non-reporting of injuries because of the benefits that can be derived in the form of prizes, awards, trophies, etc. Injury-free does not necessarily mean a high degree of safety control, nor does it mean accident-free.

Activities around injuries are mostly post-contact. Correct injury analysis and proper use of the information could assist in predicting and preventing further undesired events, which could lead to similar injuries.

A more accurate measure of safety is required and the measurement of certain elements of safety controls and efforts to reduce potential of accidental loss on an ongoing basis would prove to be far more substantial. No safety system or safety program can guarantee a reduction in the number of injuries due to the three luck factors. Measuring safety controls that help reduce the likelihood and that drive the risk into the As Low As Is Reasonably Practical (ALARP) area, are far more pro-active and accurate measurements of safety.

CHAPTER ELEVEN

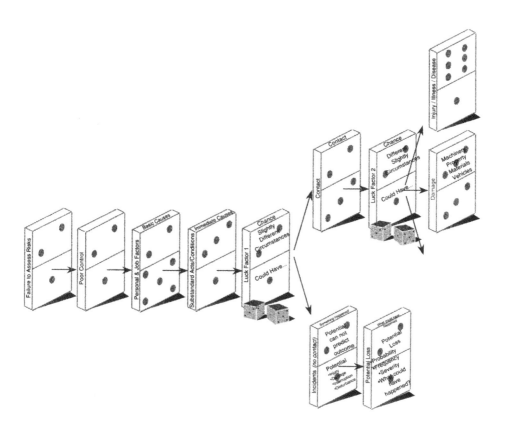

PROPERTY AND EQUIPMENT DAMAGE

The exchange of energy caused by an unsafe act, unsafe condition, or combination thereof could result in injury, damage to property and equipment, or business interruption.

There are numerous accidents that result in damage to property or equipment. This damage could extend to raw materials, vehicles, the environment or finished products. Fire and explosions caused by accident are perhaps the most devastating types of property-damage events.

The damage is usually as a result of contact and exchange of energy greater than the resistance of the item. The environment can be damaged as a result of both fire and pollution and extensive losses can occur even when no injuries take place.

Most property-damage accidents also result in business interruption and some form of financial loss.

Property Damage

Property-damage accidents are the most important in the accident ratio. They are warnings that a failure exists in the management system. The property-damage accident often happens as a result of motorized vehicles colliding with buildings, the raw product, or the finished goods.

Property-damage accidents also have the potential to injure people and therefore should not be ignored. All property-damage accidents should be thoroughly investigated and a costing done to calculate the actual financial losses as a result of the accident.

Cost of repairs to equipment and vehicles should also be listed and tabled at the various safety committee meetings. These statistics form a vital part of loss statistics.

Costing

Bird and Germain (1992) list a few major types of property damage and these include:

- buildings

- fixed equipment

- motor vehicles

- tools

- materials

- materials handling equipment (p.73)

They also give the various classifications of property damage as:

1. minor (less than $100)

2. serious ($100 to $1,000)

3. major ($1,000 to $10,000)

4. catastrophic (over $10,000)

Example

The Flixborough disaster of 1974 has been termed Britain's greatest industrial accident. In addition to the 29 people who died in the explosion, more than 100 were injured and 100 homes in the village nearby were destroyed or badly damaged.

> The blast took off slates and whole roofs in the village of Flixborough itself; all the windows were shattered; doors wrenched off and walls cracked. Chimney pots came tumbling down into the street and people were hurled about like rag dolls. Within moments, the peaceful scene resembled something out of the wartime blitz, said John Kennin, a witness.

Further damage was caused as a result of the explosion, as it totally destroyed Nypro's eighteen million pound (sterling) Flixborough plant, which was reduced to rubble as a result of the explosion.

Accident Ratio

In referring to Frank Bird's accident ratio there are, according to Bird and Germain (1992) 30 property-damage accidents for every accident that results in serious injury. This means that only 2.5% of all accidents result in injury, and 73% in property or equipment damage.

Loss Control

The term "loss control" was first coined by Frank Bird in his book *Damage Control,* where he stated that bridging the gap between traditional injury prevention programs and loss-control programs meant the recognition, investigation, and reduction of accidents that resulted in property and equipment damage. In shifting the focus from the tip of the iceberg i.e., the injuries, Bird and Germain (1994) explain the concept:

> Third, if the event results in property damage or process loss alone, and no injury, it is still an accident. Often, of course, accidents result in harm to people, property and process. However, there are many more property-

damage accidents than injury accidents. Not only is property damage expensive, but also damaged tools, machinery and equipment often lead to further accidents. (p.18)

They continued their motivation of the importance of the property-damage accident by explaining that investigating of property-damage accidents could lead to the basic causes of the event that, under different circumstances, might have led to injury of personnel.

In a newsletter of the National Safety Management Society, the reference to Frank Bird's concept is explained as follows:

> Mr. Bird pioneered in the expansion of industrial safety from an injury-oriented concept to a discipline encompassing all accidents by his extensive studies and writing on the identification, costs and control of all property damage accidents during the 1950's and early 60's. The book – *Damage Control*, co-authored by him and published by the American Management Association in 1966 was one of his many publications on this subject. (*Insight*, January 1993 – Convention Special)

Writing in the same newsletter, William C. Pope, P.E., CSP, referred to a series of studies that Bird and Germain conducted in 1954 and he states that they "bridged the gap between accident and total accident prevention."

> All too often companies failed to recognize the extent and frequency of industrial accidents. Usually only personal injury accidents are reported, investigated, and analyzed; and frequently others are ignored as near-misses despite the fact that they indicate potential hazards and cost thousand of dollars in property damage alone. (*Insight* – January 1993)

The success that Frank Bird and George Germain (1992) had in introducing the concept of loss control is explained in *Practical Loss Control Leadership*:

> During this period *damage control* provided a logical bridge from injury-oriented safety programs to accident-oriented programs. More and more people recognized not only that accidental damage is extremely expensive, but also that damage accidents have significant potential for injuring and killing people. *Damage Control* is the first book published on a totally new approach to plant safety that places the emphasis on *all* accidents – not just those resulting in injuries. Today damage control is recognized as a vital part of safety/loss control by leading organizations around the world. (p.8)

Bird and Germain (1992) further explain why property damage accidents were seldom considered:

> As we considered a ratio, we observed that 30 property damage accidents were reported for each serious or disabling injury. Property damage accidents cost billions of dollars annually and yet they are frequently misnamed and referred to as "near-accidents." Ironically, this line of thinking recognizes the fact that each property damage situation could probably have resulted in personal injury. This term is a holdover from earlier training and misconceptions that led supervisors to relate the term "accident" only to injuries. (p.21)

Costs

Accident Facts, 1993 edition, published by the National Safety Council, estimates that motor vehicle damage for the year 1992 was in excess of US$38 billion. The same publication quotes the cost of fire losses, which include fires to vehicles, outside storage, crops, and timber as being more than US$10 billion for the same period. (p.3)

Even in medium-sized organizations, property damage as a result of accidents should be tracked and costed to establish a true cost of these accidents. An example of property-damage costs caused by accidents is given below:

January 1995	$6,930
February 1995	$47,114
March 1995	$32,795
April 1995	$5,968
May 1995	$108,656
June 1995	$21,587
July 1995	$8,709
August 1995	$19,124
September 1995	$3,006
October 1995	$14,775

November 1995 $8,580

December 1995 $250

Total for 1995 $277,495

The above figures are taken from actual property damages reported by a company and give some indication of the extent of property damage that could be incurred by accidents.[4]

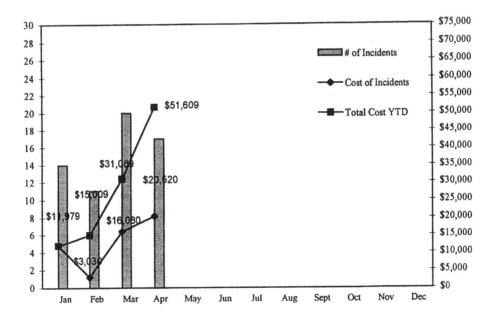

Model 11.1 - Shows the monthly cost of property-damage accidents, the total number of accidents and the progressive costs.

Importance

As emphasized by Bird and Germain (1992), property-damage accidents are caused by exactly the same sequence as injury-producing accidents. As they also have potential to cause injury, they should receive the same attention and investigation as injury-producing events. To determine the total cost of risk within an organization, the costing of all losses as a result of undesired events should be done monthly, and on a progressive basis. This information should be circulated to management and departmental management, and should be discussed at safety committee meetings. Although an organization may have relatively few injuries, their property-damage accidents may be costing a great deal.

FAILURE TO ASSESS RISKS	LACK OF CONTROL	BASIC CAUSE ANALYSIS	IMMEDIATE CAUSES	CONTACTS (Energy)	LOSS and Costs (Estimated Loss)	

• Decline risk.
• Maintenance
• Vacation planning
• Production pressure
• Training weaknesses
• Barrier operation
• Ongoing coaching
• Safety awareness of electricians
	• Non compliance to standard reporting and correction of equipment in bad order.
	• Inadequate training program standards which generalize training for different equipment of the same type.
	• Non-compliance to standard reporting damage and operation of barrier
	• No risk assessment standards.
		Personal Factors:
		• Driver was exposed to undue demands resulting in stress. (Supervisor's expectations)
		• Improper motivation to save time and effort.
		Tow Motor skill:
		  • Inadequate initial instruction.
		  • Inadequate practice. (#5 Motor)
		  • Infrequent performance of task. (
		  • Lack of coaching. (Tow Motor)
		Job Factors:
		• Management does not plan for replacement equipment, in the event of mechanical break down, resulting in Electricians desire to keep damaged equipment to get job(s) done.
		• Inadequate employee performance measurement and evaluation. (Proper use of barrier)
		• Inadequate assessment of maintenance needs and engineering evaluation of changes. (Normal practice to invert cross arm brake adjustment bar)
		• Inconsistent use of standards, policies and rules combined with inadequate monitoring of standards compliance.(Reporting damages. – electricians and dispatcher)
		• Inadequate work planning or programming. (Vacation, replacement personnel)
		• Inadequate assessment of loss exposure – engineering. (2% slope)
			Unsafe Acts:
			• Driver operating equipment in poor state.
			• # 5 Battery Motor was not tagged in poor condition , and in use.
			• Joseph (Keith's partner) demand to operate equipment on bad order.
			• Mechanics inverting brake adjustment cross arm assembly, resulting in assembly bolts facing up.
			• #5 Battery Motor equipment check card not properly filled out.
			Unsafe Conditions:
			• # 5 Battery Motor left in unsafe condition, not locked out.
			• Poor communication between "C" shift and "A" shift workers.
			• Decline barrier was left in improper position.
			• Battery qualification detonates setting on the brake assembly bolts, resulting in inoperable brakes.
			• #3 Tow Motor missed scheduled maintenance. (Mechanic on maternity leave)
			• Tow always include "Tow Motor".
			• 2% slope.
				1) Driver struck against cab of #5 Battery Motor.
				2) #5 Tow Motor struck against other carriages.
				3) #5 Tow Motor and carriage fell to a lower level.
					Injury to employee.
					
					Property damage to #
					5 Tow Motor.
					
					Property damage to motor and carriages.
					
					5 Hours of production lost.
					
					5 Hours of supplying wire goods lost.
					
					5 Hours of Electricians use of battery motor, lost
					
					Accident Investigation Costs:
					• Electricians
					• Crew
					• Supervision
					• Contractors Management & Safety
					   Management & Safety
						$28,000
						
						10,890
						
						3,000
						
						1307
						
						500
						
						200
						
						1000
						
						$44,987

Model 11.2 - The CECAL model used to investigate a property-damage accident.

The damage listed in the above model was to the tow motor, the carriages, and the electrical system. Various items can be damaged during one accident and thorough investigation techniques should identify all exchanges of energy and all subsequent damages.

Examples

Some examples of property damage accidents reported are given:

> The employee was passing by when he saw another employee in the lunchroom dozing off. When he hit the window to get his attention the window broke.

I stopped the truck and got off to direct the mechanic on the forklift and I thought I had put the gearshift into park. The truck rolled and ran into a flatbed trailer.

The riggers were checking the cables on the auxiliary lift and the block was lowered to the ground. Excessive cable was unwound off the drum and this came into contact with an energized electric welding lead lying on the ground. This burnt a portion of the auxiliary hoist's cable.

A crane knocked the handrail off the deck and pulled out the electrical cable for the controls of the mixing machine.

A forklift truck was stuck. A second forklift driver picked up the back of the stalled forklift truck and got his forks too far under it and put a hole in the oil pan.

The crane driver was going to the south end of the building to get a drink of water. He thought that the crane was in the neutral position and when he approached the end of the building he looked down and noticed the control was in the drive position. He moved the control back but there was not enough time. The crane hit the wall.

These property-damage accidents clearly indicate the potential for injury and also the costs of repairing the accidental damage as well as the accompanying disruption.

Non-Reporting

The same reasons for the non-reporting of injuries exist for property-damage accidents. Top of the list is a fear of disciplinary measures as property damage accidents are expensive and result in disruption. A good safety system should encourage all property damages to be reported. A costing of the accident should be done and any damage in excess of $1,000 deserves a thorough investigation and attempt to prevent a recurrence of a similar type of accident. The events leading up to a property-damage accident should be investigated as if it were an injury-causing accident.

Conclusion

Accidental losses could include losses to people, equipment, material, or the environment. In Madagascar, an accidental oil spill polluted some of its most famous beaches but, because of a lack of financial resources, these areas could not be reclaimed and remained polluted for a number of years adversely affecting the country's tourist industry and destroying natural habitat. Property-damage accidents are a vital part of a safety program and should be investigated, costed, and treated with the same amount of diligence as injury-producing accidents.

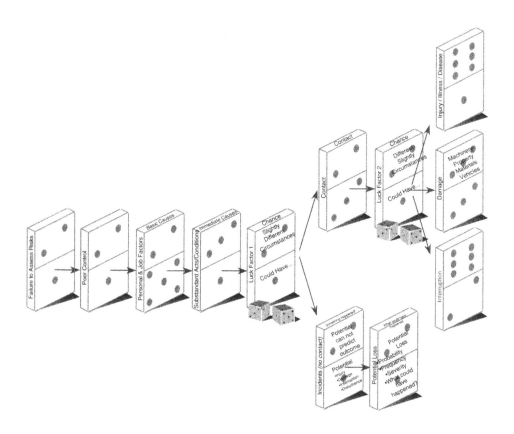

CHAPTER TWELVE

BUSINESS INTERRUPTION

An *accident* has already been defined as *"an undesired event that results in harm to people, damage to property or business interruption."* Business interruptions are perhaps the most intangible consequence of accidents and are seldom treated as true accidents.

Luck Factor 2

As already explained, Luck Factor 2 determines which of the three categories of loss will be produced by an accidental exchange or contact with a source of energy. They are: injury or disease, property and equipment damage, or business interruption, or a combination of either two or all three types of loss. Injuries, occupational diseases, and property damage tend to receive more attention than business interruptions or loss of processes. Loss of process or business interruption is normally accepted as part of doing business even though the events leading up to the business interruption are identical to those leading up to injury or property damage. Traditionally, there has been very little concern shown for or recognition of the fact that business interruption and loss of process result from an accident.

All Accidents cause Loss

All accidents, whether they result in injury, property damage or business interruption, cause a loss. An interruption of normal day-to-day business can also be caused by an accident. Some accidents however, do not cause injury or result in any damage whatsoever, but stop or retard the intended outputs of the organization and thus interrupt or hinder the process.

This interruption could be in the form of:

- lost time

- lost product

- loss of process

- poor quality

- rework

These business interruptions as a result of accidents are seldom, if ever, reported and investigated and consequently very seldom is an effort made to reduce their causes. Frank Bird introduced the concept of Loss Control and helped shift focus from the injury to property damage as a more frequent, and just as important, consequence of an accident. A further thrust is needed in the future to broaden the focus to include the loss of product and business interruptions caused by the same series of events that caused other losses.

As with basic cause analysis, investigation and costing of business interruptions due to unplanned events could be like opening a can of worms, as these interruptions clearly indicate poor management control and supervision.

Loss Control

Bird and Germain (1992) feel that loss control is a part of every manager's job and give three reasons for this:

1. Managers are responsible for the safety and health of others.

2. Managing safety provides significant opportunities for managing costs.

3. Safety/loss control management provides an operational strategy to improve overall management. (p.41)

In their book *Practical Loss Control Leadership* they further quote Dr. W. Edwards Demming as a renowned quality consultant who states that only 15% of a company's problems can be controlled by employees – while 85% can be controlled only by management. "In other words most safety problems are management problems." (p.41)

Operational Problems

Since the management system is responsible for budgeted and predicted output of an organization, any interruption would reflect directly on a deviation from intended management activity.

Dan Petersen (1978) also linked accidents to weaknesses in the management system:

> The root cause of accidents (weaknesses in the management system) is also the cause of other operational problems. Though this fact is not immediately obvious, the more we consider it, the more obvious it seems to become. Consider, for instance, how often our safety problems stem from lack of training – and how often our quality problems also stem from the same lack of training. Or consider how poor selection of employees creates safety problems and other management problems. The fundamental root causes of accidents are also fundamental root causes of many other management and operational problems.

This is not a new thought. Heinrich (1959) expressed it in a slightly different way: "Methods of most value in accident prevention are analogous with the methods for the control of quality, cost, and quantity of production." (p.18)

Quality

The sequence of events that normally affects the quality of the process as well as the quality of the end product is very similar to the sequence that results in injury to people. Failing to assess the risk of undesired events occurring in the planning, production, and marketing of products could also result in poor quality and expensive product recalls required by law.

A product recall caused by an undesired event similar to an accident is perhaps the most expensive business interruption that could be experienced by an organization.

The Consumer Product Safety Commission (CPSC) requires that manufacturers recall all goods or products manufactured and sold that pose a hazard to the end user. Incorrect wording of the installation or usage instructions, use of faulty materials, or failure to test the integrity and duration of the product sufficiently can easily be related to the cause and effects currently being discussed.

Examples of product recalls indicate that an undesired event (unsafe product) progressed right up to Luck Factor 1. One assumes that there has not yet been a member of the public injured by the product. The product has potential for loss that could occur under slightly different circumstances.

Examples

As reported in the March 1998, edition of *Professional Safety*, "Hedstrom Corporation, Bedford, PA, has recalled $1.5 million glide rides, which were sold with backyard gym sets. The unit's j-bolt assembly can break, causing children to fall."

In another product recall *Professional Safety* (March 1998) quotes: "Toys "R" Us has recalled 4,000 children's craft sets because of a potential fire hazard. The manual incorrectly instructs users to microwave soap discs for 10 minutes instead of 10 seconds." (p.11)

Another product recall is as follows: "Retailers nationwide have recalled 1.5 million sets of curtain-style indoor/outdoor holiday lights. Wiring can easily detach from splices, exposing live wires and presenting an electrical shock hazard." (p.11)

Product recalls as a result of poor quality and failure to thoroughly assess the risk of the product by adequate pre-marketing inspection and testing are perhaps the costliest of interruptions.

Worker Morale

Simonds and Grimaldi (1963) discuss the peace of mind of the worker being affected as a result of accidents, which could lead to slowdown in the work throughput.

> In a no-injury accident the factors of unpleasantness and possible temporary physical impairment would not be so cogent, but even then, according to the definition, there must be an element of danger. Therefore, even no-injury accidents are not likely to be conducive to the peace of mind of the workers. (pp.86-87)

This means that a slowing down or extra caution exercise by workers as a result of an accident could start to hinder the normal production. They continue discussing the costs related to accidents as follows:

> If production lost because of an accident is made up by working overtime, the accident should be charged with the difference between the cost of doing the work in overtime and the cost that would have been incurred if it had been done in regular hours. (p.88)

Here they are referring to interruptions or delays and time lost as a result of an interruption caused by an accident. An interruption could also have a chain reaction and affect other workstations, especially in production-line-type industries.

Simonds and Grimaldi (1963) continue their accident-cost discussion by including workers who are forced into temporary idleness by the accident but are still being paid their normal wage. They also include their cost of supervision having to investigate and rectify the business interruption as one of the hidden costs of business interruptions.

Simonds and Grimaldi (1963) include other costs connected with business interruptions and continue:

> Among such possible costs are public contracts canceled or orders lost if the accident causes a net long-run reduction in total sales, loss of bonuses by company, cost of hiring new employees, if the additional hiring expense is significant, cost of excessive spoilage (above normal) by new employees and demurrage. (p.91)

Although they were referring to injury-producing accidents with the accompanying interruption and consequent costs, the same argument could be held to stand for the accident that only causes interruption, and, due to fortuity, does not cause injury or significant property or material damage.

Putting forward a very strong argument concerning the costing of accidents they continue:

> When workers or machines are made idle, one of two things occurs. Either the production is eventually made up, or it is not. If it is made up, in the sense that over a long period of time no less goods are produced and sold than would have been had the accident not occurred, there is no loss of profit, apart from the increase in production cost. (p.93)

They also include the non-availability of an organization by using an example of a contractor:

> If a contractor in the construction industry has some accidents that necessitate spending an extra 3 weeks on a given job, this may reduce the number of contracts he is able to undertake that year. Manufacturers in a war or immediate postwar period often are able to sell all they can produce. Thus, in some instances accidents cause a net reduction in production and sales. When this occurs, some "loss of profit" is a legitimate charge. If a production delay enables competitors to secure some of the firm's business, there is loss of profit. (p.93)

To link a loss of profits to the interruption or loss of process caused by the accident they then summarize as follows:

1. Accidents must cause a decrease in average rate of output over a considerable period of time.

2. The result and rate of output must be lower than management finds it desirable to maintain in view of the demand for its product and average and variable unit cost of production.

3. This resultant rate of output must be sufficiently lower to cause a reduction in sales, due to inability to supply the goods.

4. Sales lost during this period must not be recoverable at a later date.

Summing up, the following conditions must all be present in order to create a "loss of profit" on goods not produced as a result of accidents. (p.94)

Increased or Decreased Productivity

In a three year study on the impact of a good safety program on productivity, Bird and Germain (1992) quote the following statistics:

> As a result of the introduction of a safety management system, the mine experienced:
>
> - a total accident frequency reduction of 57%
> - the disabling injury frequency reduced by 87%
> - labor turnover rate was reduced by 57%
> - plant availability increased by 13%
> - production output increased by 40%
> - tons per employee increased by 15%
> - machinery availability increased by 4%.
> - degree of safety on the total mine was up 27%. (figure 1-7, p.12)

Bird and Germain (1992) also state:

> A substantial drop in accident injury frequencies with corresponding savings in costs, time and production shifts, as well as improved employee morale all attributed to the implementation of the program. It is proved positively that safety and productivity go hand in hand. (p.12)

On quoting H.W. Heinrich, Bird and Germain (1992) repeat Heinrich's statement by saying: "He wrote, methods of most value in accident prevention are analogous with methods required for the control of the quality, cost and quantity of production."

Bird and Germain (1992) continue linking the management of productivity with the management of safety by saying:

> Without exaggeration, one of the biggest needs throughout the world among organizations that have not attained the desired levels of loss control performance, is this matter of managing the work necessary to get the desired job done efficiently and effectively. (p.42)

Business Interruption

A *business interruption* is defined as *an undesired event that does not result in injury nor damage but which produces an undesirable change to the normal flow of process or manufacturing, thus interrupting the business.*

If one couples this definition with the Bird and Germain (1992) definition of *loss* which is *"avoidable waste of any resource,"* (p.42) it becomes clearer that focusing on accidents that interrupt the business could identify weaknesses in the management system and rectification measures could in turn improve both quality, throughput, and production.

Example

In an organization that relies on tonnage produced, a study was undertaken and it was found that approximately 10% of the process was lost as a result of accidents.

An analysis of the loss of product was undertaken and it was discovered that 24% of the product was being lost due to spills, 10% lost to mechanical failures and 7% as the result of motor failure. A further 30% was lost as a result of miscellaneous causes. Further investigation showed that 90% of the causes for the lost process were accidental and could have been prevented. Model 12.1 shows the percentage of process lost as well as a breakdown as to the causes of 10% of the production being lost.

An analysis was taken over a six month period and the results are produced in Model 12.1

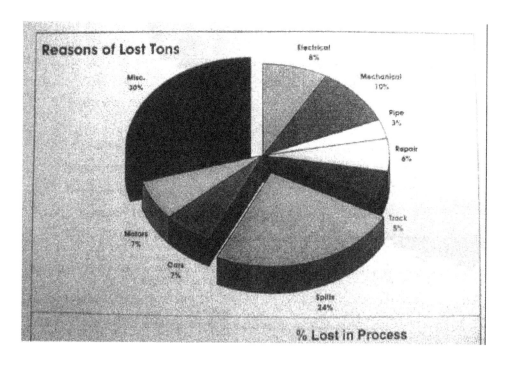

Model 12.1- Shows the total percentage of process lost per month.

At least assessing the risk and instituting control measures can prevent 98% of all undesired events. This means those business interruptions, loss of process, and poor quality can be reduced and controlled in the same way as undesired events that result in injury.

Heinrich et al. (1969) quotes the Goals Freedom Alertness Theory as follows:

> Dr. Willard Kerr has proposed a theory that states that great freedom, to set reasonable attainable goals, is accompanied typically by high quality work performance. The theory regards an accident merely as low-quality work behavior – a "scrappage" that happens to a person instead of to a thing. The more rich the climate in reward opportunities, the higher the levels of alertness, the higher the levels of work quality, and the lower the probability of an accident. (p.44)

Conclusion

An accident can therefore result in one or a combination of three major consequences.

1. occupational injury or illness

2. property or equipment damage

3. business interruption in the form of loss of process, low productivity, or poor quality

The business interruption as a result of an accident receives less attention than spectacular property damage and injury to people, yet occurs more frequently than the latter. In relating to the Bird ratio, if there are 30 property-damage accidents for every serious injury, there must be at least 100 accidents that disrupt the normal output of an organization.

Business interruptions could mean workers are idle, machines are not fully utilized, time is wasted, the number of units produced is reduced or the intended production flow is hampered.

Business interruption accidents should be costed and their causes identified and rectified. As with property-damage accidents, business interruption accidents do have potential to cause damage and/or injury. Production management should include the avoidance of business interruption accidents in the same way as accidents that cause injuries to employees.

CHAPTER THIRTEEN

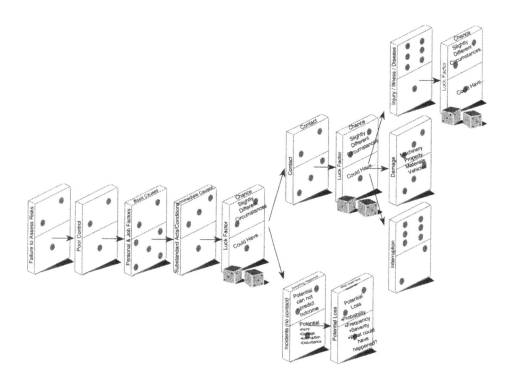

LUCK FACTOR 3

Heinrich et al. (1969) compiled 10 axioms of industrial safety, the most pertinent one to this chapter being axiom no. 4, which states *"the severity* of an injury is largely fortuitous – *the occurrence* of the accident that results in injury is largely preventable." (p.21)

This fourth axiom is perhaps the most significant statement in the safety management profession. What Heinrich is saying here is that the degree of injury depends on luck, but that the accident can be prevented. What he further indicates by this axiom is that while the accident can be prevented, the severity is something over which we have little or no control.

Luck Factor 3

In examining the Cause, Effect, and Control of Accidental Loss sequence (CECAL), once an injury occurs as a result of an exchange of energy, the degree of injury is largely dependent on Luck Factor 3. Most safety activities are focused around the severity of an injury or illness. Consequently, the focus is on an end result, which is determined by fortune, chance, or luck.

Even in the fourth edition of *Industrial Accident Prevention*, printed in 1959, H.W. Heinrich and E.R. Granniss further explain Luck Factor 3 by stating, "There are certain types of accidents, of course, where the probability of serious injury may vary in accordance with circumstances." (p.28) In other words, the circumstances that determine the probability of a serious injury are really those of either good or bad fortune.

They continue their discussion as follows:

> An injury is merely the result of an accident. The accident is controllable. The severity or cost of an injury that results when an accident occurs is difficult to control. It depends upon many uncertain and largely unregulated factors – such as the physical or mental condition of the injured person, the weight, size, shape, or material of the object causing the injury, the portion of the body injured, etc. Therefore, attention should be directed to accidents as properly defined, rather than to the injuries that they cause. (pp.28-29)

Pattern

All loss-causation events follow the CECAL pattern but their progress through the loss-causation model is channeled either by Luck Factor 1, Luck Factor 2 or Luck Factor 3. The difference between a fatality, permanent disabling injury, temporary disabling injury, lost-time injury, and a first aid case is largely a matter of luck.

In examining hundreds of cases where the degree of injury could have been far greater than it was, it is difficult to explain the resultant injury other than by conceding it was luck, as depicted by Luck Factor 3.

Example

The following example shows how the degree of injury is fortuitous.

Electrical Contact Accident: Because the electrician assumed the contacts were de-energized, he did not check with the meter to verify the situation, and he did not take hot-work precautions. Using a wire cup-wheel in an electric drill, he was cleaning one contact when he apparently bridged between two of the three contacts initiating an arc that, once started, bridged all three phases. A witness stated that the arc flash enveloped the electrician. The arc caused burns to his wrist and neck. It also burned holes in his long-sleeved work shirt, scorching it all the way up both arms to the back of his neck; pitted his safety glasses; singed facial hair and hair on the back of his neck. It scorched papers protruding from his shirt pocket. The arc tripped a breaker, cutting power to the plant. The electrician was treated at the clinic, released back to his normal duties and completed his scheduled work shift.

Because the electrician was not seriously injured, this was not recorded as a reportable injury or a lost-time injury and the safety "record" was not affected at all. Reading the description of the accident it appears obvious that to all intents and purposes the electrician should have been electrocuted, yet suffered only a minor injury. S.H. Hawkeswood, pr. Eng. (1977) says that any current between 16 and 1,500 milli-amps can cause "various effects such as severe pain, paralysis, and ventricular fibrillation." He then gives a case history:

> A young girl, who suffered from a heart ailment, picked up a badly joined flexible lead of a bedside lamp with one hand. She was standing on a carpet and the burns on her hand indicated contact between the line and neutral at her hand only. Although the power was isolated within a very short time, the girl died. (Case History 6)

The two accident scenarios clearly indicate that the degree of injury is difficult to predict and is therefore determined by other circumstances beyond our control.

More Shocks

In analyzing a minor injury log, I came across four situations where the accident type was "contact with a source of electricity." Of these four different accidents, none resulted in serious injury and no one was electrocuted or severely burnt.[2]

Accident 1: "The pipe fell, slipped forward and cut the 480-volt electrical cable, shocking some men."

Accident 2: "I was climbing the ladder near the crane gantry and I felt an electrical shock."

Accident 3: "There was a 440-volt cable on the ground. I placed some walk boards down to kneel on them and grabbed the steel with my right hand while my left hand brushed the cable. I got shocked."

Accident 4: "I was changing the position of the stepladder, which hit an electrical cable and shocked me."

Having contact with medium to high voltage and not experiencing a fatality is indeed luck. Ventricular fibrillation can occur from 15 milli-amps upwards. The young girl who died as a result of a relatively insignificant contact with electricity was less fortunate.

Dan Petersen (1997) says, "Quit looking at accident-based measurements to assess systems effectiveness." (p.40) What Dan Petersen is referring to is that, because of Luck Factor 3, any measurement of safety performance based on degree of consequence is based on degree of *luck*. Admittedly, the number of serious injuries and fatalities is important, as is the number of first aid and dressing cases. So much emphasis is placed on *lost-time, disabling* injuries and *reportable* injuries that this degree of harm to the body has become the focus of most safety programs, safety practitioners, unions, and regulatory bodies. If one accepts Heinrich and the other safety pioneers' statement that the degree of injury is fortuitous, then using the degree of injury as a measurement tool is really measuring an organization's *luck factor*. Even boards of directors, when quoting lost-time or disabling-injury rates in annual reports, regard these measures (of safety) as being "internationally accepted measurements of safety performance." They are really measurements based on the degree of consequence after three factors of luck that have determined the outcome. From the unsafe act or unsafe condition, Luck Factor 1 determines whether there is going to be an exchange of damaging energy. Once there is an exchange of damaging energy, Luck Factor 2 determines whether that energy will cause a business interruption, property damage, or personal injury/illness, or a combination of two or all three. Once that energy is directed toward a human body, Luck Factor 3 (or circumstances beyond our control and our prediction) determines how much energy is going to be exchanged with what part of the body, under what circumstance. The result could be anything from a Band-Aid wound to a fatal injury. Dan Petersen is absolutely correct when he advises not to use the degree of consequence as a measure. James Tye openly stated that frequency rates were measurements of failure rather than success. Dan Petersen was also quoted in an article in *Occupational Hazards* in January 1997 as saying "You go from an occupational injury rate of 6 to 4 in one year, does this mean that you are x percent better? If so, what is it that you did to get better? Organizations usually say: 'I have no idea. Just lucky, I guess.' " (p.41)

W. Tarrants (1980) also said that:

> To the extent that it can be argued that all injury-producing events are the consequence of the same causal structure and that the degree of injury is merely a matter of chance, there are clear advantages in a measure that includes all events in the sample. (p.180)

He continues by saying, "At any future point in time, accidents that have a potential for loss may produce a loss. Which exposure results in loss and the degree of severity of future losses are influenced by chance factors." (p.15)

What Tarrants is referring to in this statement is the three luck factors. These determine whether there is going to be a loss or a near-miss. If there is a contact, luck determines whether there is going to be injury or property damage and disruption or, if there is going to be injury, how severe it will be.

In emphasizing how Luck Factor 3 clouds management's vision as to the effectiveness of the safety system, Tarrants (1980) explains: "One big problem is the accident (injury) that escapes detection because it appears below the fine dividing line established by the ANSI disabling injury or the BLS-OSHA recordable injury or illness definitions."

He continues by explaining why injury rates are not a good form of safety measurement:

> None of the measures of safety performance commonly in use is acceptable as a valid measure of identifying our internal accident problems. It is important that one is not lulled into believing that present measures are really descriptive of the actual level of safety effectiveness within an organization. (p.40)

Continuing the justification of not using categories of injuries as a measurement of safety, Tarrants (1980) explains:

> A measurement problem exists in that low injury frequency rates tell nothing about the potential for catastrophe. For example, one plant experienced a $5 million accident loss, with its walls covered with safety awards received based on a low frequency rate. Better measurement techniques are needed to identify loss potential.

He continues his argument:

> Rates are too crude. They fluctuate too much, they are too unstable, and changes in the rate do not necessarily reflect changes in the level of safety within the organization measured. The rate can fluctuate drastically on a fortuitous basis. Likewise, measures of severity can fluctuate widely by chance. For example, a person can fall and bruise an elbow. One can also fall from the same height and fracture the skull and die. The cause can be the same, but the effects and ultimate results are substantially different. (p.239)

Here Tarrants is once again emphasizing that the injury and severity are fickle measures of performance because of the chance outcomes as depicted by Luck Factor 3.

In a recent study undertaken by the Accident Prevention Advisory Unit of the Health and Safety Executive, of the United Kingdom, the extensive study was summarized as follows: -

> Any simple measurement of performance in terms of accident (injury) frequency rates or accident/incident rate is not seen as a reliable guide to the safety performance of an undertaking. The report finds there is no clear correlation between such measurements and the work conditions, in injury potential, or the severity of injuries that have occurred. A need exists for more accurate measurements so that a better assessment can be made of efforts to control foreseeable losses.

Safety Paradigm

The case against using the degree of severity of an injury as a "safety measure," is certainly strengthening. As has been seen, safety pioneers acknowledge that the severity of an injury is fortuitous. They agree that a large factor of luck exists in determining the resultant injury. Yet safety practitioners, regulatory agents, and most management gauge their safety on the number of injuries that lost a shift or longer. On achieving one million injury-free man-hours, massive safety celebrations are held at factories, plants, and mines all around the world. Yet it has never been scientifically proven that the introduction of a safety-management control system has a direct bearing on the number or severity of the injuries. This was also stated by the Health and Safety Executive's intensive report. What is certain is that assessing the risk and instituting controls reduces the probability of unsafe acts and unsafe conditions occurring but because of the three luck factors, injuries *will* occur and, depending on luck, they may be disabling, minor, or even fatal. The challenge to the safety fraternity, and to management, is to stop regarding the count of a certain level of injury as a measure of good or poor safety. The degree of control, the activities, and ongoing safety-management system are a far better measurement of safety than the number of lost-time, reportable or disabling injuries, which are determined by luck.

Dan Petersen (1978) strengthens the argument by saying:

> Perhaps our biggest doubt comes when we ask ourselves whether a death count or death rate really reflects our national effort and performance

in safety. We seem to believe that fatalities are caused largely by "luck," and thus we tend to raise our eyebrows at such a measure. (p.128)

He continues to explain Luck Factor 3 as follows:

> It measures a level of failure somewhat less than a fatality (an injury serious enough to result in a specific amount of time loss from work), but the fact remains that the supervisor of 10 workers can do absolutely nothing for a year and attain a zero frequency rate with only a small bit of luck. By rewarding such a supervisor, we are actually reinforcing non-performance in safety. (p.66)

Krause (1997) refers to the accident triangle and explains: "An at-risk behavior whose outcome lies off of the triangle is an incident. An identical behavior whose result lies on the triangle is an accident. Behavior based safety investigators are not confused by the difference in chance outcomes." (p.293)

His triangle has at the base at-risk behavior, followed by four levels of injury to the apex of the triangle. The first level is first aid followed by recordable, followed by lost-time, and finally by a fatality.

Questionnaire

In a written survey involving 23 employees, question no. 11 reads as follows: Do you believe that "luck" has a part in the outcome of an accident and the severity of an injury. Of the 23 replies all replied "yes" to both parts of the question.

Case Study

In this particular fatal accident, the degree of severity was worsened by luck. The following factors were listed that could have worsened the degree of injury.

1. The victim was only released from the cab of the damaged vehicle 47 minutes later. The length of time to free the victim may have played a role in allowing the severity of injuries to increase.

2. The light was fading, which may have hampered the rescue operations.

3. Traveling time to the closest medical center was 2 hrs. If a medical rescue helicopter had been summoned as soon as the accident happened it could have been on site at the time of the victim's release from the cab.

4. If the vehicle had fallen on its other side, the degree of injury might have been less.

Conclusion

The type of injury that results from an accident is largely dependent on factors that can neither be predicted nor controlled, and which numerous safety authors refer to as fortune or chance. H.W. Heinrich's (1959) one-line statement in *Industrial Accident Prevention* seems to have been grossly overlooked by the safety profession. They have ignored what he has said and based all efforts, recognitions, punitive measures, successes and failures on the degree of injury as a result of an accident. Numerous studies have also concluded that there is little, if any, correlation between good safety controls, a good safety program, and the number and severity of injuries. The Middelbult Coal Mine disaster in 1987 is a prime example of this. Having been rated a five-star mine by NOSA for three years preceding the accident, it had also maintained a disabling-injury incidence rate of less than one for three years concurrently when, in one underground explosion, 53 miners were killed.

Using the degree of injury as a measurement of safety performance or safety failure is completely misleading. The difference between minor injuries, disabling injuries, lost-time, and fatal injuries is largely a matter of luck.

CHAPTER FOURTEEN

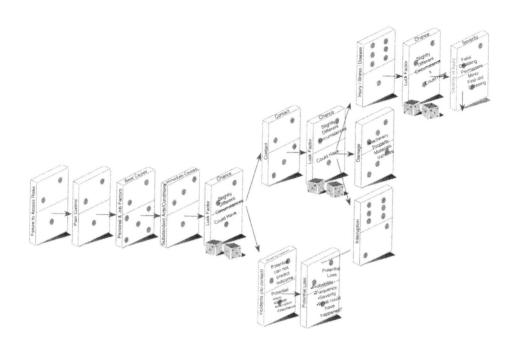

SEVERITY OF INJURY

The severity of an injury is largely dependent on Luck Factor 3 and many safety pioneers and previous safety management research confirm this fact. The degree of injury could range from a minor abrasion requiring only a plaster to a fatal injury. Depending on the amount of energy transferred, the area of body involved and the duration of the energy exchanged, Luck Factor 3 determines how badly the person or persons are about to be injured.

Definition

The *severity* of the injury could be defined as *the degree of injury or illness suffered and the extent of the disability of the person that was injured.*

The Auditor's Guide, NOSA (1995) defines an *injury* as *"any damage to people usually caused by a single short exposure of an individual to the high energy exchange between two or more sources of energy."* (p.20)

NOSA further defines a *disease* as *" any damage to people most usually caused by the frequent, continuous, extensive or long exposure of an individual to one or more sources of energy."* Its definition of *disability* is *"the inability or limited ability to perform a pre-injury activity in the same manner or within the same range as was possible prior to the damage."* (p.20)

Attraction

The injury, and more specifically the extent of injury, is what is traditionally linked closest to the terms "safety" and "accident." Most people focus on injuries and the focus is intensified by the severity of the injury. A fatal injury is perhaps the most dreaded experience that management could have, yet the energy exchange that killed the person could have, under slightly different circumstances, inflicted only minor injury or injury less severe than a fatal injury.

The number of serious injuries measures most safety programs and safety successes. Because so much focus and emphasis is put on the severity of injury, safety practitioners throughout the world have often been inclined to focus their attention on the injured person in an effort to ensure that the injury is not regarded as "measurable" or "reportable." Certain injuries must be reported by law and these injuries are also selected by their severity. The most popular significant injury being the "lost-time injury," or the more modern version "the disabling-injury." All injuries are in fact "lost-time injuries" but what the term intends is that the person loses a complete work shift, other than the shift on which he was injured, as a result of the severity of the injury. The *disabling-injury* definition is similar to the "lost-time" except it is more explicit and states that the injured employee is *temporarily partially or totally disabled for a shift other than the shift on which he was injured.*

Survey

In the 15thAnnual White Paper survey conducted by the *Industrial Safety and Health News* (December 1998), management was asked how safety and health specialists could contribute to their companies. Ninety one percent answered by saying, "reduced injuries and illnesses." A further 76% wanted reduced Worker's

Compensation costs. None indicated that they wanted a reduction in the risks but rather indicated they wanted a reduction in the degree of consequence.

Heinrich (1959) emphasized that accidents are what are important:

> **Accidents - not injuries - the point of attack.** Accident prevention is too frequently based upon an analysis of the causes leading to a major injury. This situation exists, for the most part, because of a misunderstanding of what an accident really is.

He continues by saying:

> In reality, where the term "accident" and "injury" are so merged, it is assumed that no accident is of serious importance unless it produces a serious injury. Yet thousands of accidents having the potentiality of producing serious injuries do not so result. (p.28)

Safety Measurement

Few organizations measure their "safety" other than by the number of days that they have managed to work without a lost-time injury or by the number of injuries per million man-hours worked. These are all measures of consequence. If the degree of consequence is fortuitous, as explained in the previous chapter, what is really being measured is luck and not control.

Some types of measurement, based on injuries are:

1. days without a fatality

2. fatality rates

3. lost-time-injury frequency rates

4. disabling-injury incidence rates

5. injury severity rates

6. total injury rates

7. reportable-injury rates

8. injury and severity indexes

All of these measurements are based on *degree* of injury and do not necessarily tell an organization how good its safety controls are.

Dan Petersen (1998) explains:

> In the early days of safety, accident (injury) measures, such as the number of accidents, frequency rates, severity rates and dollar costs, were used to measure progress (of a corporation, location or unit). Practitioners felt comfortable using those measures even though they did not offer much help – these measures did not reveal whether the safety system was working; diagnose what was/was not working; nor indicate whether the system was in or out of control.

> Although it became clear long ago that these measures offered little help, they continue to be used today. Are accident (injury) data useful for anything? It would save much time, effort and money to simply answer "no," and focus on measures with meaning. (p.37)

Safety and health legislation around the world also require injuries, based on the extent of the injury, to be recorded and reported. Different countries have different rules and criteria for reporting and therefore any attempt to compare reportable, recordable or lost-time injuries among countries is virtually impossible. The only accurate comparison is the fatality rate, which is the only injury rate that cannot be modified, adjusted, or slanted. The rules and extent of a fatality are universal.

A low injury rate does not necessarily mean good safety performance. An injury rate is not safety performance but rather safety *failure*. Safety audits that measure and compare control mechanisms introduced to prevent the cause and effect of accidental losses, should rather be used as a gauge of safety performance or safety control.

Petersen said that, even today, companies issue awards, plaques, stickers, and prizes for individuals and groups being injury free for certain periods. When asked what these individuals or groups did differently to remain injury free, the answer was that they did nothing. They were more than likely injury free as a result of luck and not control and therefore, these awards were being awarded for their having been lucky.

As Tarrants (1980) said:

> No doubt the most reliable index we have is the number of deaths. One simply cannot hide the body under the rug and forget about it. As the injury decreases in severity, it becomes progressively easier to ignore it or to remove it from the "reportable" or "recordable" category. For example, when pressure is put on the supervisor to cut down on his first aid cases, he may tell his employees not to report their minor injuries to the dispensary, but to see him for some antiseptic and an adhesive bandage.

> None of the measures of safety performance commonly used is acceptable as a valid means of identifying our internal accident problem. It is important that one is not lulled into believing that present measures are really descriptive of the actual level of safety effectiveness within an organization. (pp.39-40)

Simonds and Grimaldi (1963) feel that they are good measures and are useful for comparison:

> *Frequency* and *severity* are accepted standards by which a company can appraise its industrial injury record and set goals for achievement. Very roughly, these terms refer, respectively, to the *relative* frequency of occurrence of major injuries, on the one hand, and the total days lost, plus time charge for deaths and permanent impairments resulting from major injuries, on the other hand. (p.31)

Annual reports as well as monthly reports of large industrial and mining undertakings now frequently include impressive bar charts and graphs showing the 12-month running disabling-injury incidence rate or lost-time-injury frequency rate. Used as the sole gauge of safety performance, they are unreliable indicators of safety control.

Using the injury-severity rate, which is the number of shifts lost as a result of injuries, (which is also dependent on the circumstances of the accident that caused the injury), is not a reliable measure of good safety controls either. Dan Petersen explains (1978):

> Safety professionals for years have been attacking frequency in the belief that severity would be reduced as a by-product. As a result, our frequency rates nationwide have been reduced much more than our severity rates have. One state reported a 33% reduction in all accidents

between 1965 and 1975, while during the same period the number of permanent partial disability injuries actually increased. (p.19)

Definitions

In defining what an organization regards as a lost-time injury, a disabling injury or even a work-related injury or illness leads to increased confusion. This leads to further inaccurate reporting. The definitions differ from organization to organization and from country to country. Comparing injury rates based on different sets of rules gives a totally inaccurate picture. The American National Standards Institute (ANSI) gives the most comprehensive and rigid definitions of injuries in its standard on recording and measuring work injury experience, (ANZI Z16.1 and Z16.2). Even with these guidelines, Workers' Compensation rules differ from state to state and country to country, and some organizations accept a compensatable injury as a work injury even though it may have occurred off the premises and was not arising out of and during the course of normal employment. These differences once again detract from the accuracy of defining the degree of injury.

Group Study

Five teams, consisting of five employees each, were asked what a disabling-injury incidence rate of 8 meant.[3] Only one of the teams got the answer close to a plausible explanation of what the magical figure of 8 meant. They answered by saying, "For every 200,000 man-hours there are eight disabling injuries." This was the closest answer received from the teams. Past experience has also shown that at least 90% of the workforce do not understand figures quoted depicting disabling-injury incidence rates or lost-time injury frequency rates. Most of the teams questioned tried to recall the formula and one team actually stated, "The DiiR is measured by incidents or accidents occurring in a specific time frame, total man-hours worked divided into the reportables for the period."

The main reason for using the disabling-injury incidence rate instead of the lost-time-injury frequency rate is that the disabling-injury incidence rate is based on 100 men working for a year. This is equivalent to a percentage of the work force injured (disabling injuries) per year. This is a measurement with which the work force can associate. The lost-time injury frequency rate, on the other hand, is calculated around a million man-hours, or 500 people working for a year, which is more difficult to comprehend than percentage of work force injured. From discussions, interviews, and tests with employees in numerous industries over the last 26 years, it is clear that not only is the measurement of safety suspect but it is also *the* magical number. These are displayed on safety publicity boards around the companies, and are not understood. This is perhaps a reason that the modern trend is to display the number of days worked without suffering disabling injury. This is a more understandable indicator for the general work force.

Since Heinrich's first accident ratio was published, other researchers show that a large number of minor injuries are experienced in comparison with the number of serious or disabling injuries. The first Heinrich ratio estimated that there were 29 injuries for every one serious injury. The Bird and Germain Ratios produced in 1966 predicted 10 minor injuries for every serious injury. If an organization uses the injury at the peak of the triangle to measure its safety performance, surely the 10 minor injuries should also be taken into account. But for the luck factor, they could also have been disabling injuries. The ratios produced by Heinrich, Bird, Tye, and Pierson all indicate that the disabling injury is preceded by numerous less-severe injuries, which, under slightly different circumstances, could have been more serious, thus heightening the argument that measurements based on severity are really measuring a degree of luck.

Incentives

The debate about safety incentives continues. Some management theories advocate the recognition of good performance by giving out incentives. Safety management is no different, except that most incentive and awards are based on a measurement compiled using injuries or injury-free periods. These incentives, awards and competitions are referred to as "typical safety-incentive programs."

Mark A. Friend (1997) says that sometimes these incentive programs do more harm than good:

The operative word here is "typical." Typical programs reward employees for hiding facts – by encouraging them to not report injuries. Conversely, sound safety incentive programs reward employees who offer input that helps everyone operate more safely. (p.36)

Safety incentives do reduce the number of lost-time injuries or whatever other degree of injury is used to measure so-called safety "performance" or "improvement." In a survey conducted (1998) among 23 employees, 13 or 57% agreed that they would not report an injury if they were paid a monthly bonus of $150 for being "injury free." They unanimously agreed that if the injury was so severe that it could no longer be hidden, or that it obviously affected their ability to work, only then would they report the injury.

Judy Prickett (1998) quotes the following example:

> I recently heard of a company that promised $100 cash to all employees if they exceeded their previous record of days without a lost-time incident (injury). Within days of attaining the record, one of the employees was injured and reported that injury. He was ostracized by his fellow employees and ended up quitting. (p.8)

So, does a safety-incentive scheme really reduce the number of reportable injuries or does it merely drive those injuries underground? If the degree of injury is the sole measure of safety performance, it can easily be modified to achieve whatever number management deems as being reasonable, or a figure that shows an improvement on previous figures. *Industrial Safety and Hygiene News* (September 1998) asked for feedback from readers as to how much pressure safety departments face to under-report injuries and illnesses. This is to avoid OSHA targeting them now that the agency is using employers' lost workday rates to compile lists of inspection targets. An anonymous reply was as follows:

> There are many organizations that will never make safety a priority. They believe they are safe because no one has been killed or seriously injured recently. These folks will probably go to great lengths to fudge the numbers. We will always need a strong OSHA to deal with them! (p.21)

Another respondent to the same question in the same article starts his reply with:

> Under reporting would depend on to whom the safety department reported. If they were part of a human resources department, then the pressure would not be as great as it would be if they report to a profit center. There is always pressure to lower an injury rate, but when budgets and bonuses are on the line, it can be intense. (p.21)

From the above comments it is apparent that there is tremendous pressure to lower injury rates by any means other than to institute proper controls based on risks identified by the risk assessment process. Dr Mike Manning (1998) cautions the safety practitioner as follows:

> I have been tempted to make exceptions to the rule for the incident rate, but I never did. I always required that the injury be recorded by the department that it occurred in. If you start making exceptions for certain cases, the whole intent of the incident rate, as a measuring tool will be lost. Stand firm on this. Some supervisors will attempt to manipulate, cajole, threaten and whine their way out of accepting their responsibility to their employees, but don't let them. (p.36)

Honesty

Honesty is therefore required for both the recording and reporting of all injuries and more specifically those that eventually meet the criteria as the *attention getter* or measuring injury such as the disabling or lost-time variety. It should be mentioned that all undesired events "lose time" and therefore the use of the expression "lost-time injury" is outdated as all injuries *also* lose time. Industries, management and safety professionals should start using correct terminology and revert to using the term "disabling injuries" as defined by the ANSI Z16 standard.

If one therefore summarizes the conditions needed to have meaningful injury statistics, they should be honest and unbiased. Even though they are unbiased, are all injuries being reported? If the rules for determining disabling injuries are being followed, whose rules are these and how many exceptions have been made to them?

The intent of the disabling-injury definition was to classify injuries that actually hampered the business process, where the work in progress at the time of the undesired event could not be continued exactly as before, due to the undesired event (accident). The only acceptable definition of a disabling injury is to ask, "What was the person doing at the time of the accident?" And, "Can he return to continue the same task that he was doing after the accident without losing a complete shift?" If one were to use the above criteria to determine whether an injury is disabling, the disabling-injury-incidence rate, I believe, would double.

Dan Petersen (1996) says:

> In the early days of safety it was simple – we just looked at the statistics; a 0.5 accident (injury) frequency rate meant we were good and that our safety program was effective, whereas a 15.0 frequency rate meant that our program was not effective. It was simple; the frequency rate (of recordables or lost times) told us whether or not everything was working. And this was true until we started looking at the real truth – at the meaninglessness of the number we are using. (p.15)

The Health and Safety Executive of the United Kingdom carried out an extensive safety study via its Accident Prevention Advisory Unit and summarized this as follows:

> Any simple measurement of performance in terms of accident (injury) frequency rates or accident/incident rate is not seen as a reliable guide to the safety performance of an undertaking. The report finds there is no clear correlation between such measurements and the work conditions, in injury potential, or the severity of injuries that have occurred. A need exists for more accurate measurements so that a better assessment can be made of efforts to control foreseeable risks.

Can Safety Systems Help?

It has always been believed, and many safety practitioners still believe, that an intensive, management-driven, safety program will reduce the number of injuries. They believe this even though studies such as those done by the *Health and Safety Executive* found that often there is no correlation between the degree of control being exercised by an organization and the number of disabling injuries being experienced. Even with five-star safety programs regularly audited and acclaimed by external auditors, disasters can occur. In 1988, the Middelbult Colliery had maintained a five-Star NOSA Safety Grading for three years. This entailed maintaining 95% control of 73 crucial safety elements and a 12-monthly progressive disabling-injury-incidence rate of less than 1% per annum. Even with all this effort, an underground methane explosion occurred and 53 miners died instantly. Was this perhaps as a result of the three luck factors that have been discussed? Safety programs and control systems can only drive down the *probability of* undesired events occurring, but the actual end result is attributable to the three luck factors. The degree of an injury as well as the severity of an injury are fortuitous and therefore using them as sole measures of safety efforts is fruitless.

Example

To further emphasize the futility of using the degree of injury as a measure of safety or non-safety, the following case study is given.

An employee was working on the 65-ft level, and entered across Section 16. As he reached up to hook his fall-protection line to the static line, he slipped and fell through the rails. He fell 120 ft to the ground level, fracturing his left ankle. Further injury was averted by the build-up of material on the side of the uprights that slowed his descent. This was not recorded as a lost-time or disabling injury as the employee was back at work the following morning even though he was walking with the aid of crutches and his ankle had been tightly bandaged. It was recorded as a minor injury. The irony is that an identical accident occurred some two years previously, where the employee fell the same distance and was killed. That accident was reported as a fatal accident, and besides shocking the work force and the local community, resulted in an investigation by the local sheriff, the state safety organizations as well as federal organizations. The only difference in the two scenarios is Luck Factor 3, which, simply said, means that *luck* determines the severity of the injury.

Conclusion

The degree of injury is as a result of certain factors referred to in this publication as Luck Factors 1, 2, and 3. As a result of three luck factors, the severity of an injury is therefore a poor gauge of either success or failure in safety control systems.

Safety incentives over the years have taught employees, supervision management, and indeed, the safety profession, how to manipulate the figures, bend the rules, and produce low injury rates to justify their existence or their supervisory skills. Regulatory authorities have added to this dilemma by basing harsh inspections on companies with injury rates higher than the national injury rates for that type of industry. In other words, the honest organization that methodically reports and records the number and severity of its injuries is likely to be punished by being subject to an inspection and subsequent citations. Employees themselves are encouraged not to report injuries. If no injuries of a serious nature occur, management turns a blind eye and thinks that all is well and that it has control. Many a management team has been absolutely staggered by a fatality. In many instances, the potential for a fatality had been present all along. The same event probably occurred without severe result many times in the past.

Dave Johnson (1998), editor of *Industrial Safety and Hygiene*, quotes Henry Lick, who is a committee member and manager of industrial hygiene for the Ford Motor Company: "We don't get progress until we kill people. It's too bad progress has to be measured in blood. Whole generations go by without problems being solved." (p.4)

CHAPTER FIFTEEN

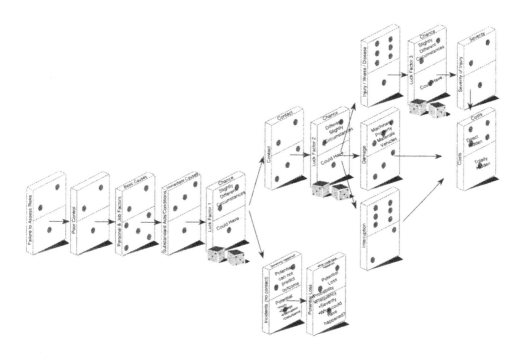

COSTS OF ACCIDENTAL LOSS

Admiral Ben Moreel (1948), president of Jones and Laughlin Steel Corporation said:

> Although safe and healthy working conditions can be justified on a cold dollar and cents basis, I prefer to justify them on the basic principle that it is the right thing to do. In discussing safety in industrial operations, I have often heard it stated that the cost of adequate health and safety measures would be prohibitive and that "we can't afford it." My answer to that is quite simple and quite direct. It is this: "If we can't afford safety, we can't afford to be in business."

Accident Costs

Accident Facts (1993) gives the total costs of accidents to the United States' economy as follows: "Besides the estimated $399.0 billion in economic losses from accidents in 1993, loss of quality of life from these accidents is valued at an additional $790.1 billion, making the comprehensive cost of accidents $1,189.1 billion in 1992." (p.2)

The end result of the CECAL sequence can always be translated into costs. Whether the event results in injury, disease, and damage to machinery, property or materials or business interruption, they all cost the organization money. Traditionally, these costs have been tolerated as the cost of doing business and have not received management's full attention. Workers' Compensation normally covers the direct costs of the accident. These premiums can be increased as a result of injury costs and these increases could remain for three-year cycles or periods. Most of the accidental costs are hidden costs or indirect costs. These are difficult to calculate and are often ignored because they do not create an immediate financial drain on the organization.

Heinrich et al. (1969) delved further than the direct costs of accidents and wrote about the incidental costs of accidents:

> In addition there are the very real so-called incidental cost of accidents. These incidental costs of accidents have been estimated to be four times as great as the actual costs. There have been innumerable studies made, discussions held, articles written, and arguments presented about that figure. No one has disputed the concept, the fact that there are indirect, or incidental or hidden costs surrounding accidents. All tend to agree with that. There is tremendous disagreement, however, about how much those costs might amount to. We might first briefly look at what the costs involved are, and then look at some of the controversy of how much they really amount to. (p.82)

In listing some 11 types of hidden accident costs, he excludes compensation and liability claims, medical and hospital costs, insurance premiums, and costs of lost time except when actually paid by the employer without reimbursement. The 11 main hidden costs he lists are:

1. cost of lost time of injured employee

2. cost of time lost by other employees who stopped work

3. cost of time lost by foremen, supervisors, or other executives

4. cost of time spent on the case by first aid attendant and hospital department staff who were not paid for by the insurance carrier

5. cost due to damage to the machine, tools, or other property, or to the spoilage of material

6. incidental costs due to interference with production, failure to fill orders on time, loss of bonuses, payment of forfeits, and other similar causes

7. cost to employer under employee welfare and benefit systems

8. cost to employer in continuing the wages of the injured employee in full

9. costs due to the loss of profit on the injured employee's productivity and on idle machines

10. costs that occur in consequence of the excitement and worker morale due to the accident

11. overhead costs per injured employee (light, heat, rent, and other such items, which continue while the injured employee is a non-producer)(p.83)

The list of costs as a result of an accident can be a lot longer depending on the circumstances. The safety pioneers established that there is a definite ratio between the insured (direct) and uninsured (indirect) costs of accidents and *conservatively* put the ratio at 1:5. A further category can be added to the costs of an accident. Those costs are termed the totally hidden costs. These costs are effects that cannot be costed out accurately and include such intangible costs as the cost of pain and suffering of the injured employee, the disruption of the victim's family life, the lowered standard of living due to his pay packet being reduced substantially, as well as the accompanied stress and anxiety caused by the accident. The loss to the community and the organization is also difficult to compute.

Iceberg Effect

McKinnon (1995) refers to the iceberg effect:

> The iceberg effect is where the majority of accident costs are hidden below the waterline. The hidden costs or indirect costs of accidents include items such as damage, loss of time and production and business interruptions. Experts have put these costs at anywhere from 5 to 50 times more than the direct costs. Irrespective of what figures are placed there is definitely a higher cost of hidden costs than direct costs.

> All profits are recuperated from the profits of a company. Peter Drucker, the well-known management consultant and author, states that it would be better to *minimize* losses than to constantly endeavor to *maximize* profits. Management is often aware of the efforts to maximize profits by improving quality and productivity, but is not always aware of losses occurring as a result of accidents. The prevention of all downgrading accidents is a good investment and can improve the company's bottom line substantially. (p.31)

Bird and Germain (1992) summarize the advantage of safety control by saying, "Save a dollar in accident costs and you add a dollar to profit." (p.22) They continue by saying, "Add this to the overriding human element and you have the best of both worlds: protection of profits, process property and PEOPLE. This is why it is so essential to understand and use the accident cause and effect sequence."

They also list the ratio between the injury and illness costs and the ledger costs of property damage as anywhere between 5 to 1 and 50 to 1. "The uninsured miscellaneous costs are quoted as 1 to 3 dollars per every 1 dollar direct cost." (p.24)

In motivating the amount of profit consumed by accident costs, Bird and Germain (1992) give an excellent example: "If an organization's profit margin is 5%, it would have to make sales of $500,000 to pay for $25,000 worth of losses. With a 1% margin, $10,000,000 of sales will be necessary to pay for $100,000 of the costs involved with accidents." (p.25)

Main Motivation

For years, safety practitioners and safety organizations have promoted the humanitarian aspect of safety and have endeavored to reduce injury-producing accidents. The humanitarian approach is all very well, but this is not what motivates management. Management is in business to make a profit and give shareholders a return on their investment. Profits and the bottom line are what an organization is all about and what management at all levels is ultimately held accountable for. If the costs of accidents due to poor control can be brought to management's attention, the necessary actions and support will be forthcoming. Yet, so many safety people concentrate on the numbers of injuries, and endeavor by all means to reduce these numbers, that they lose sight of their greatest management attention getter, *the total cost of accidents*. Simonds and Grimaldi (1963) stated:

> The second major objective of accident prevention work is to reduce production or operating costs for the sake of profit. While second to the prevention of human injury, since that is even more directly important as a human value, cost reduction broadens the basis for safety work. Cost reduction provides a direct purpose for preventing all kinds of accidents, those that happen not to cause injury as well as those that do, and cost reduction as a purpose brings into focus the losses from property damage and interference with production as well as the purely injury aspects. (p.29)

Individuals, promoters, and organizations geared to prevent accidental loss will continue to be unsuccessful unless they start to justify their recommendations on sound financial basis.

In an interview with Jerry Scannell, president, National Safety Council (October 1998), *Compliance Magazine* asked the question, "What are some new initiatives at NSC?" Scannell replied:

> In the workplace, we are trying to emphasize the relationship between safety programs and cutting costs. This fall, we will be holding a symposium with the Japan Industrial Safety and Health Association that will promote studies that explain the relationship between effective safety programs and saving companies money. (p.2)

Compliance Magazine then asked, "How can a company achieve excellence in safety?" Scannell replied, "Success will not happen overnight. A company must hold safety as a core value right up there with making a profit."

In answering the question, "How are you convincing management to promote safety within their companies," Scannell answered, "If you are trying to convince your company's chief executive officer that safety is important, you must show that safety offers financial rewards. 'Cut costs' are two words that corporate America is using today." (p.2)

In emphasizing the importance of hidden costs, Dan Petersen (1978) says, "Hidden costs are real." (p.118) In discussing the measurement of safety consequence he says, "Many people today believe that the dollar is a far better measuring stick than any other in safety, and many companies are beginning to utilize it effectively." (p.50)

It should be remembered here that if the severity of injury is as a result of Luck Factor 3, then the subsequent direct costs of the injury or disease are also largely fortuitous and therefore although the dollar cost is a useful measurement, it, like injury rates, is highly subjective.

Fines

Most organizations are required by legislation to maintain a safety program, which gives certain controls of risks that have been identified. Although these controls, checks, and balances do cost money, they should be viewed as an investment, as accidental losses could be far more expensive. Non-compliance to legislation could also cost a company a staggering amount of money. *Compliance Magazine* (October 1998) gives some examples of the cost of non-conformance to basic legal standards:

1. An airfreight corporation in Ohio agreed to pay $250,000 in penalties and make significant improvements in safety and health programs at their Vandalia Hub at the Dayton International Airport.

2. For alleged exposure violation penalties were assessed totaling $123,900.

3. An organization agreed to pay OSHA $90,000 for willful and repeated violations of failure to provide a workplace free from recognized hazards.

4. Another organization was fined $30,000 for willful and serious violations as a result of an inspection that was conducted following an inadequate response by the employer to a complaint.

5. $163,800 was paid for willful violations involving failure to secure a scaffold that exposed employees to falls of up to 25 ft.

6. Bypassing a safety limit switch and using a wooden stepladder on top of a lift work platform, cost an organization in Ohio $138,600. (pp.17-18)

Highest Ever

Compliance Magazine also quotes the largest penalty ever assessed against a mine as being $1,250,250.00, which the mine agreed to pay including penalties, assessments, and costs to settle civil and criminal proceedings. "This is the largest penalty assessed against a metal and non-metal mine and should serve as a reminder to everyone in the underground mining community that vigilance in maintaining safety precautions is a vitally important manner," said labor Secretary Alec Herman. (p.26)

Cost Benefit

Dr. Mike Manning's (1997) approach to the safety problem is supported by many other practitioners who do this cost benefit analysis during the evaluation phase of a risk assessment. Thorough risk assessment will indicate to management where best to spend its money for the largest return. Traditionally, a number of safety programs have channeled money into activities that have not necessarily tackled the root cause of the accident problems.

Tarrants (1980) supports this as follows:

> When various loss prevention approaches are being considered, one basis for decision is to examine a cost of each versus the benefit of its application. The alternate countermeasures that produced the greatest benefits for the least cost are normally selected. One problem with this approach is the difficulty in assessing dollar costs of such intangible

accident correlates as pain, suffering, and loss of life. What is life worth? (p.258)

Life's Value

In answering Tarrants' question, the National Safety Council (1992) gives the costs of an accidental death as $780,000 for the year 1992. It reports the cost per disabling injury as $27,000, which includes loss of wages, medical expenses, administrative expenses, and employer costs, but excludes property-damage costs except to motor vehicles. (p.35)

Purdey (1995) quotes David Rainor, director of safety, Railtrack, on their decision not to install automatic train protection on the British Rail network as required after the Clapham Junction inquiry, as follows:

> Automatic train protection would be horrendously expensive to install, at a cost of 14 million Pounds (Sterling) for every life saved. No final decision has been made, but we are beginning to conclude that there is not even a case for partial installation. (p.419)

Purdey (1995) continues by saying:

> A suitable valuation of statistical fatalities for decision making purposes is tentatively suggested to be in the region of 2 million Pounds (Sterling)(1991 prices). More safety conscious companies would be justified in using valuations up to 50 million Pounds. (p.420)

In referring to the above quotations it is obvious that costs both of non-conformance and of total conformance is the deciding factor. One of the largest penalties paid by a South African division of an international organization was when, in 1994, four chemical firms agreed to pay 20 South African workers R9,4 million (US$1.5 million) in damages and costs.

Reputation

One of the hidden costs of an accident is the cost a company suffers when its reputation is tarnished as a result of a serious accident or series of accidents. On the day following the Vaal Reefs disaster on 17 May 1995, the shares of Anglo

American Gold Division, the owners of Vaal Reefs, fell R16 (US$3) per share in one day.

Even ignoring the intangible costs, an accident that results in a fatality normally has severe repercussions on the organization. In one particular fatal investigation, the costs were broken up into injury, property damage, business interruption, and total hidden costs. The summary of this accident report was "the losses of the accident are listed under the headings of injury, property damage, interruptions, and total hidden costs. Although not costed out, total cost is estimated to be far in excess of US $1,000,000."

Even property-damage accidents can hamper the production of an organization and one particular property-damage accident, which resulted in a seven-day period to repair the damage, resulted in a loss of 5,000 units being produced.

White Paper

Industrial Safety and Health News' 15th Annual White Paper survey (December 1998) asked management to list the most important contribution safety and health personnel can make to their companies. According to the White Paper only 37% of the managers listed, "Document the financial impacts of safety activity." When asked what skills and knowledge are important for a safety professional, 17% responded, "The ability to document dollar savings of safety activity." (p.22) It was also interesting to note from the survey that almost half (44%) of the managers surveyed believed accidents will happen, regardless of safety efforts. (p.22)

Total Cost of Risk

There are three main areas where costs play a role in safety. The first is the cost of the end result of undesired events such as injury and property-damage accidents. As discussed, these costs could have as many as three different tiers and can range from tangible to intangible and from direct to indirect costs. They are losses to the organization nevertheless. The second cost of risks is the cost to insure equipment, plant, product, and personnel. These are the Workers' Compensation costs, insurance costs, and such like. The third cost is that involved in reducing, containing, and minimizing the risks, which manifests in loss and cost-producing events. This is the cost of the safety program and personnel.

Summary

The CECAL sequence always ends up with costs as the last effect. Once the event results in a contact the losses could be due to injury, property damage, business interruption, or a combination of all three as already discussed. The costs of assessing and controlling the risk have proved to be less expensive than the cost of the consequence of the event. The benefit of reducing and controlling the risk is also the avoidance of heavy legal fines for non-compliance to legal safety standards. The costs of safety controls are therefore a good investment and as one safety professional put it, "If you think safety is expensive, try an accident!"

REFERENCES

Bird, F. E. Jr. and Germain, G. L. (1992). *Practical Loss Control Leadership* (2nd ed.). Loganville, Georgia: International Loss Control Institute.

Bird, F. E. Jr. and Germain, G. L. (1996). *Practical Loss Control Leadership* (3rd ed.). Loganville, Georgia: Det Norske Veritas.

Boylston, R. P. (1990). *Managing Safety and Health Programs.* New York: Van Nostrand. © Reprinted by permission of Whiley-Liss, Inc., a subsidiary of John Wiley & Sons, Inc.

Carroll, Lewis. (1978). *Alice's Adventures in Wonderland.* London: Octopus Books.

CM. (1998). Airline cited for forklift accident. *Compliance Magazine,* 18. © Reprinted with permission of *Compliance Magazine,* IHS Publishing Group. All rights reserved.

CM. (1998). Corporation pleads guilty to mine safety violations. *Compliance Magazine,* 6. © Reprinted with permission of *Compliance Magazine,* IHS Publishing Group. All rights reserved.

CM. (1998). Legal briefs. *Compliance Magazine,* 17. © Reprinted with permission of *Compliance Magazine,* IHS Publishing Group. All rights reserved.

CM. (1998). Speaking with Jerry Scannell. *Compliance Magazine,* 2. © Reprinted with permission of *Compliance Magazine,* IHS Publishing Group. All rights reserved.

Consumer Product Safety Commission Product recalls. (1998). Information originally published in *Professional Safety,* 11-12.

Daniels, A. C. (1994). *Bringing out the Best in People.* New York: McGraw-Hill, Inc.

De Ionno, P. and Dlamini, J. (1995, May 14). The level 72 horror. *Sunday Times,* p. NNEWS9. Johannesburg.

Ferry, T. S. and Weaver, D. A. (1976). *Directions in Safety.* Springfield: Charles C. Thomas.

Friend, M. A. (1997, Feb.) Examine your safety philosophy. *Professional Safety,* Feb., 34-36.

Geller, E. S. (1996). *Working Safe.* Radnor: Chilton Book Company.

Grose, V. L. (1987). *Managing Risk.* New Jersey: Prentice-Hall.

Heinrich, H. W., Petersen, D., and Roos, N. (1969). *Industrial Accident Prevention* (5th ed.). New York: McGraw-Hill Book Company.

Heinrich, H. W. (1959). *Industrial Accident Prevention* (4th ed.). New York: McGraw-Hill Book Company.

Howe, J. (1998). A union view of behavioral safety. *Industrial Safety and Hygiene News,* 20.

Hudson, L. A. *Insights into Management.* Safety Management Society, 2. An internal document of Western Deep Levels mine entitled, *A Guide to Effective Accident/Incident Investigation.*

International Loss Control Institute. *International Safety Rating System* (6th ed.). Loganville: International Loss Control Institute.

Johnson, D. (1998, September). Cooking the books. *Industrial Safety and Hygiene News, 21.*

Johnson, D. (1998). OSHA strains to serve its many masters. *Industrial Safety and Hygiene News, 7.*

Johnson, D. (1998). What does management think about safety and health? *Industrial Safety and Health News:* 15th White Paper survey, 22.

Krause, T. R. (1997). *The Behavior-Based Safety Process* (2nd ed.). New York: Van Nostrand Reinhold. © Reprinted by permission of Whiley-Liss, Inc., a subsidiary of John Wiley & Sons, Inc.

Manning, M. V. (1998). *So you're the Safety Director!* (2nd ed.). Rockville: Government Institutes, Inc.

Manuele, Fred A. (1997). *Professional Safety*, 31.

McKinnon, R. (1995). *Five Star Safety: an Introduction*. Unpublished work.

National Occupational Safety Association. (1988). *Advanced Questions and Answers*. (Vol. HB5.11E). Pretoria: National Occupational Safety Association.

National Occupational Safety Association. (1990). *Effective Accident/Incident Investigation*. (Vol. HB4.12.50E). Pretoria: National Occupational Safety Association.

National Occupational Safety Association. (1993). *MBO 5 Star Safety and Health Management System*. Pretoria: National Occupational Safety Association.

National Occupational Safety Association. (1994). *Health and Safety Training*. (Vol. HB5.31.55E). Pretoria: National Occupational Safety Association.

National Occupational Safety Association. (1995). *The NOSA 5 Star System*. (Vol. HB0.0050E). Pretoria: National Occupational Safety Association.

National Safety Council. (1993). *Accident Facts 1993*. (Vol. 1993). Itasca: National Safety Council.

NOSAdata 3.13.05, *Recording and Measuring Work Injury Experience*. Pretoria: National Occupational Safety Association.

Petersen, D. (1978). *Techniques of Safety Management* (2nd ed.). New York: McGraw-Hill Book Company.

Petersen, D. (1988). *Safety Management*. New York: McGraw-Hill Book Company.

Petersen, D. (1996). *Analyzing Safety System Effectiveness* (3rd ed.). New York: Van Nostrand Reinhold. © Reprinted by permission of Whiley-Liss, Inc., a subsidiary of John Wiley & Sons, Inc.

Petersen, D. (1997). Why safety is a 'people problem.' *Occupational Hazards*, 39-40.

Petersen, D. (1998). What measurement should we use, and why? *Professional Safety*, 37-39.

Prickett, J. (1998). Incentive programs reflect management's attitude. *Industrial Safety and Health News*, 12-13.

Rothstein, M. A. (1998). *Occupational Safety and Health Law* (4th ed.). St. Paul: West Group. Reproduced with permission.

Simonds, R. H. and Grimaldi, J. V. (1963). *Safety Management* (8th ed.). Homewood: Richard D. Irwin, Inc.

Smit, E. and Morgan, N. I. (1996). *Contemporary Issues in Strategic Management* (1st ed.). Pretoria: Kagiso Tertiary.

Smith, S. L. (1994). Near-misses: safety in the shadows. *Occupational Hazards*, 33-36.

Smith, T. A. (1998). What's wrong with safety incentives? *Professional Safety*, 44.

Tarrants, W. E. (1980). *The Measurement of Safety Performance*. New York: Garland STPM Press.

NOTES

1. A 15-question questionnaire completed by 23 employees attending a safety course asking whether injury reporting was honest and if they would report an injury if their bonus was lost as a result.
2. Actual accident reports, which have been changed to preserve the identity of the person and organization.
3. Questions posed to five teams of five members each during an 8-hour safety training course concerning the interpretation of "a DiiR of 8."
4. Report on a property-damage accident.
5. Risk assessment protocol form used by BHP Copper, San Manuel division, Arizona.

HOW TO INVESTIGATE AN ACCIDENT

AN INVESTIGATOR'S STEP-BY-STEP INVESTIGATION GUIDE, INCLUDING CHECKLISTS, INTERVIEW FORMS, AND BASIC CAUSE ANALYSIS GUIDES.

A Complete Accident Investigation Kit.

ACKNOWLEDGMENTS

- This booklet is dedicated to all those injured persons who were victims, not only of accidents, but also of poor accident investigations.

- My thanks to all my mentors in safety.

- My sincere appreciation to those firms and divisions that no longer invite me to investigate accidents in their area because they are afraid I might reveal the true cause of the accident and open a can of worms. This tells me I must be doing something right.

- My appreciation to Chuck Gessner, who afforded me the opportunity of investigating a number of accidents and incidents.

- Thanks to all the participants at my "Accident Investigation" Seminars for their participation and input.

- To my wife Maureen, thanks for the typing and proofreading.

Ron McKinnon, CSP

DISCLAIMER

This document does not replace any form, register, or other document required by your local safety and health legislation. Nor does it replace Insurance or Workers' Compensation claim forms as prescribed. This document does not guarantee to eliminate causes of other possible accidents or incidents. The author and publisher offer this document only as a guide to the accident investigation process and will not be held liable for any litigation arising out of the use or misuse of it whatsoever.

TABLE OF CONTENTS

INTRODUCTION

The fact that you are reading this means that you have been appointed to be responsible for investigating an accident or an incident.

To clarify the two concepts: an ***accident*** *is an undesired event that results in a loss in the form of injury to people, damage to property and equipment, or which interrupts the business process*. The keyword in an accident is that there was some form of *loss*.

An incident is also called a near-miss or a warning. *An* **incident** *is an undesired event, which under slightly different circumstances could have caused loss to people, property, or the environment*. The events that cause an incident are exactly the same as an accident. The only difference is there is *no loss*.

The rules of investigation apply equally to loss-producing events and events that did not produce a loss, but which had the potential to do so. This book can be used to investigate all undesired events.

At the outset, the golden rule of accident investigation must be stated and will be reinforced throughout the entire investigation procedure. The golden rule of accident investigation is "accident investigation is a **fact-finding** mission and not a **fault-finding** exercise."

Management Principles

In investigating accidents, there are certain management principles that need to be remembered and the first is the principle of *multiple causes*. The principle of *multiple causes* states that *accidents are seldom, if ever, the result of a single cause*. This means that during the course of your investigation you will identify more than one cause of the accident. If only one cause is uncovered during an investigation, the evidence is insufficient and the true causes will not be revealed.

The next principle to be remembered during accident investigation is the principle of *definition*, which states that *a logical and proper decision can only be made when the basic or real problem is first defined.* This means that one cannot pose solutions to a problem unless the real cause of the problem has been identified. This accident investigation kit will enable you to delve right into the root causes of the undesired event. Once these have been identified, solutions to rectify them can then be proposed. Remember that **prescription without diagnosis is malpractice.**

Legal Requirements

Please check which legal reporting and recording requirements apply to this accident, or incident investigation. The use of this document in no way exonerates, or excuses, you or your organization from any legal reporting, recording, and investigation obligations.

Cause and Effects

Before investigating an accident, one must understand how an accident occurs. An accident occurs as a result of a sequence of events very similar to the fall of a row of dominoes. W. H. Heinrich et al., proposed the first domino accident sequence in the book *Industrial Accident Prevention.* (1929)(p.22) Bird and Germain further expanded on this and although they used the basic concepts of the domino sequence, the titles of the dominoes were different from those of Heinrich's.

McKinnon further added to the domino sequence in the chapter entitled "The Strategic Importance of Safety" (p.283) in the book *Contemporary Issues in Strategic Management*, (1995) including the three luck factors that determine the course of the flow of these dominoes as well as the outcome and the severity of the outcome. In 1996, he further modified this sequence by adding the initiating domino entitled "failure to assess the risk." According to this sequence, the initiating event of all accidental loss is the failure to assess and control risks.

Accident Causes

In referring to the Cause, Effect, and Control of Accidental Loss (CECAL)© accident sequence, an accident such as the one you are about to investigate is initiated by failure to assess the risk of either behavior or work conditions. This triggers off basic causes in the form of personal and job factors, which then lead up to unsafe acts and unsafe conditions. These unsafe acts and unsafe conditions are the immediate cause of the event. Luck Factor 1 determines whether the unsafe act or unsafe condition will result in a contact with a source of energy or merely be an incident. In this case, nothing happens but there was potential for loss. Luck Factor 2 then determines the outcome of the contact with a source of energy. The outcome could be injury, illness, or disease to people, damage to machinery, property, materials, vehicles or a business or process interruption.

Injury

Should this contact unfortunately result in harm to people, Luck Factor 3 determines the severity of that harm. The severity could either be a minor injury, a more serious injury resulting in lost time, a disabling injury, a permanent disability or even a fatality. The final domino indicates that all of these adverse effects, such as injury, damage, and interruption, result in costs or a financial loss to the business.

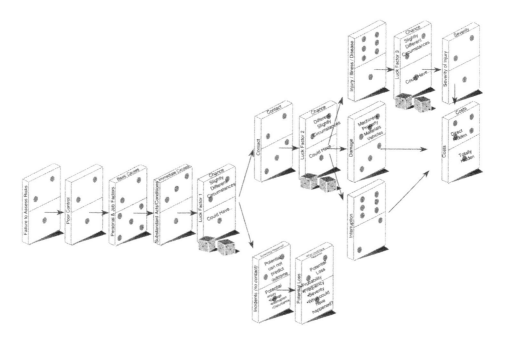

The Cause, Effect, and Control of Accidental Loss Model (CECAL)©

Road Map

The road map that you are about to follow in investigating an accident will follow each domino in sequence and gather information along the way. Eventually, the process will lead to solutions to the problems. The solutions to the problems will be in the form of safety management control measures.

Patch Prevention

Beware of patch prevention as remedies to your problem. Should a person have slipped on a puddle of oil and been injured, your recommendations should not only be to clean up the puddle of oil but to determine how the oil got there in the first place, and how to treat the real cause of the oil being on the floor, rather than the symptom.

Immediate Actions

Before an accident can even be investigated, certain immediate actions need to take place. These immediate actions are to secure the scene of the accident to prevent further accidents, injury, or damage, and administer proper first aid to the injured persons and evacuate them to a safe place where further medical treatment can be given. Do not attempt to investigate the accident until these measures have been taken, as there could be the possibility of secondary accidents causing even more injury and damage. Once the site has been secured, the area made safe and injured people taken care of, the investigation can begin. Ensure that the appropriate people and organizations are notified accordingly.

<u>STEP 1</u> - GATHER EQUIPMENT AND INSTRUMENTS

The first step in an accident investigation is to gather the necessary equipment. The equipment that you will need for a good investigation is as follows:

- Writing paper and pens (this book)

- Tape measure

- Tape recorder

- The necessary personal protective equipment for that area

- Ziploc™ plastic bags

- Camera and film

- Danger tape to mark off area

- Witness interview forms

In more sophisticated accident investigations the following additional equipment and instruments may be needed:

- Noise-level meter

- Light-intensity meter

- Wind-speed meter

- Thermometer

- Wet-and-dry thermometer

- Volt-meter

- Polarity tester.

<u>STEP 2</u> - VISIT ACCIDENT SITE, GATHER, AND RECORD INFORMATION

Accident Site

The next step is to visit the accident site. Before plunging into the investigation, walk around the site and get a feel of the conditions under which the accident occurred, taking cognizance of the various activities on the site. Then begin collecting the evidence.

Important

It is of vital importance that the accident investigation be conducted as soon as possible after the accident and that the site visit take place as soon as the immediate actions have been completed. For every hour that the site is not visited, valuable evidence is changed or disappears, and witnesses' recollections of the events start to become distorted. If the important points of an accident investigation were ranked in order, this would certainly be high on the list. Visit the site as soon as possible after the accident.

GENERAL INFORMATION

Name of Injured _____

Date of Accident _____ *Time* _____ *Location* _____

Name of Investigator _____ *Assisted by* _____

Injured Person's Payroll No. _____ *Position* _____

Supervisor _____ *Division/Unit* _____

Medical Treatment given _____

By Whom/Which Hospital _____

Other Information _____

Who should be notified?

 Supervisor _____ *Foreman* _____ *Spouse* _____

Doctor _____ *Management* _____ *Legal* _____

Insurance _____ *State* _____ *Police* _____

Ambulance _____ *Human Resources* _____

Workers' Compensation Insurance _____ *Public Relations* _____

Other Agencies _____ *Others* _____

Gather Evidence

Pictures: Pictures should be taken of the accident investigation site, of the equipment that was involved and also of the general condition of the work environment. Include positions of people, parts, and machinery.

Examine: Examine the entire area as well as adjoining areas to collect information. Examine the equipment, parts of the equipment, and also all attachments for any signs of failure, how far they travel through the area or what role they played in creating this loss.

Samples: If necessary, collect samples and keep them in Ziploc™ bags. Make sure that a sketch of the area has been done or photographs have been taken before the samples are removed. Always label the samples and indicate clearly the date and time they were collected. To enable you to do this, record the samples taken or parts removed on the evidence itinerary list as follows:

Article and/or Sample collected:

#1_____

#2_____

#3_____

#4_____

#5_____

RECORDING OF INFORMATION

A checklist for recording the conditions at the accident site should now be completed:

What was the general level of illumination in the work area?

_____

_____

_____

_____

Was the ventilation, artificial or natural, sufficient, free from odors or fumes?

What were the noise levels in the area? Record any impact noises, background noises, or noises that hampered communication.

Record the positions of people at the time of the accident.

*Person #1*_____

*Person #2*_____

*Person #3*_____

*Person #4*_____

*Other*_____

What tasks were being performed during this accident?

*Task (injured person 1)*_____

*Task (injured person 2)*_____

*Task (injured person 3)*_____

Person _____

*Other*_____

What tools were being used during the accident?

Tool 1 _____

Tool 2 _____

Tool 3 _____

Tool 4 _____

What equipment was being used?

Equipment 1 _____

Equipment 2 _____

Equipment 3 _____

Equipment 4 _____

What energy sources contributed to the accident?

❑ *Electricity* _____

❑ *Heat* _____

❑ *Chemical* _____

❑ *Gravity* _____

❑ *Steam* _____

❑ *Mechanical Force* _____

❑ *Speed* _____

❑ *Other* _____

❑ *Other* _____

❑ *Other* _____

What was the condition of the general housekeeping in the area?

Housekeeping _____

Stacking and Storage _____

Walkways Demarcated _____

Work Areas Demarcated _____

Sketch

At this stage a sketch of the area should be made.

Draw a sketch of the accident scene and indicate on the sketch the position of the injured person and/or damaged equipment and surrounding features. Be sure to include dimensions of distances, heights, speeds, etc.

Sketch of Accident Site

WITNESSES

Witnesses are vital to an accident investigation, as they were there before, during, and after the accident. They possess first-hand knowledge of what happened. Once again, it is important to interview witnesses as soon as possible after the event to ensure their recollection of the accident is still accurate. If a person (persons) was injured it is also vital to get their testimony. Ensure that they are not under duress due to pain and are in a position to give the information. If not, wait until the attending medical practitioner gives clearance for the interview. Always respect the injured person's injuries and accompanying trauma.

The following four pages provide for witnesses' accounts of the accident.

WITNESS STATEMENT

Name _____ *Position* _____

Date _____ *Time* _____

Recollection of the Accident

_____

_____

_____

_____

_____

_____

Signature _____ *Date* _____

Person interviewing Witness _____

WITNESS

```
┌─────────────────────────────────────────────────────────────┐
│ Notes:                                                       │
│                                                              │
│                                                              │
│                                                              │
│                                                              │
│                                                              │
│                                                              │
│                                                              │
└─────────────────────────────────────────────────────────────┘
```

WITNESS STATEMENT

Name _____ *Position* _____

Date _____ *Time* _____

Recollection of the Accident

_____

_____

_____

_____

_____

_____

Signature _____ *Date* _____

Person interviewing Witness _____

WITNESS

Notes:

WITNESS STATEMENT

Name _____ *Position* _____

Date _____ *Time* _____

Recollection of the Accident

_____

_____

_____

_____

_____

_____

Signature _____ *Date* _____

Person interviewing Witness _____

WITNESS

Notes:

WITNESS STATEMENT

Name _____ *Position* _____

Date _____ *Time* _____

Recollection of the Accident

_____

_____

_____

_____

_____

_____

Signature _____ *Date* _____

Person interviewing Witness _____

Accident Reconstruction

In some cases, it may be necessary to reconstruct the accident. A warning here! Be careful not to ask people to re-enact the accident so convincingly that the same event recurs and further injury and damage are incurred. Please explain this to the people involved in the reconstruction. Ask witnesses to demonstrate what was happening at the time of the accident but to do everything in slow motion and not injure themselves or others while re-enacting the accident. A record of this accident reconstruction should be made below.

Accident Reconstruction _____ *Demonstrated by* _____

Describe the Accident Reconstruction

_____

_____

_____

_____

_____

_____

_____

_____

_____

_____

_____

_____

_____

_____

_____

_____

_____

_____

_____

_____

<u>STEP 3</u> - RECORD THE TYPE OF INJURY/DAMAGE/EVENT THAT OCCURRED

Event Type

There are numerous types of undesired events. Check the box that is most appropriate and write a brief description next to the event chosen.

❑ *Near-miss* _____

❑ *Accident* _____

❑ *Property Damage* _____

❑ *Motor Vehicle* _____

❑ *Fire* _____

❑ *Industrial Injury* _____

❑ *Non-industrial Injury (not work related)* _____

❑ *Other* _____

❑ *Other* _____

❑ *Other* _____

It is very likely that the undesired event resulted in more than one of these consequences and if so, check them appropriately.

Injury Type

If there was an injury-producing accident, indicate the severity of the injury. Once again check an injury-severity box and give a brief description.

❑ *First Aid Case* _____

❑ *Hospitalization* _____

❑ *Legally Reportable* _____

❑ *Lost-Time Injury (a complete shift or more will be lost)* _____

❑ *Disabling Injury (cannot return to the task he was doing for >1 day)* _____

❑ *No Medical Treatment* _____

❑ *Other* _____

Property Damage

If the accident resulted in property damage give an indication as to the severity of the damage.

❑ *No Equipment Damage* _____

❑ *Damage <$1,000* _____

❑ *Damage >$1,000* _____

❑ *Super Damage* _____

❑ *Business Interruption* _____

❑ *Other* _____

❑ *Other* _____

<u>STEP 4</u> - DO A RISK ASSESSMENT

Now do a risk assessment of the loss potential as well as the probability of recurrence of the event. Remember, it is not only what happened that is important, but what could have happened. Use normal expected conditions.

Loss Potential (What was the loss potential?)

❑ *Low* _____

❑ *Medium-low* _____

❑ *Medium-high* _____

❑ *High* _____

Probability of recurrence (What is the possibility of it happening again?)

❑ *Low* _____

❑ *Medium-low* _____

❑ *Medium-high* _____

❑ *High* _____

Personal Protective Equipment in Use

What personal protective equipment was in use at the time of the accident?

❑ *Hard Hat* _____

❑ *Safety Shoes* _____

❑ *Respirator* _____

❑ *Gloves* _____

❑ *Fall Protection* _____

❑ *Hearing Protection* _____

❑ *Glasses* _____

❑ *Goggles* _____

❑ *Other (Specify)* _____

❑ *Other (Specify)* _____

STEP 5 - DETERMINE THE TYPE OF INJURY AND BODY PART INJURED

Type of Injury   **Explanation**

❑ *Traumatic Amputation* _____

❑ *Asphyxia* _____

❑ *Burn (heat or chemical)* _____

❑ *Contusion (bruise)* _____

❑ *Wound (laceration/abrasion)* _____

❑ *Skin Irritation* _____

❑ *Dislocation* _____

❑ *Electrical Shock* _____

❑ *Fracture (open/closed)* _____

❑ *Frost Bite* _____

❑ *Heat Exhaustion/Stroke* _____

❑ *Hernia* _____

❑ *Inflammation* _____

❑ *Sprain/Strain* _____

❑ *Multiple Injuries* _____

❑ *Puncture* _____

❑ *Foreign Object* _____

❑ *Other (Specify)* _____

❑ *Other (Specify)* _____

❑ *Other (Specify)* _____

Body Part

Once the nature of injury or type of injury has been ascertained what part of the body was affected?

❑ *Head*	❑ *Neck*	❑ *Right Eye*
❑ *Left Eye*	❑ *Mouth/Nose/Ear*	❑ *Right Arm*
❑ *Left Arm*	❑ *Back*	❑ *Upper Back*
❑ *Mid Back*	❑ *Low Back*	❑ *Chest*
❑ *Shoulders*	❑ *Right Leg*	❑ *Left Leg*
❑ *Right Knee*	❑ *Left Knee*	❑ *Right Ankle*
❑ *Left Ankle*	❑ *Right Hip*	❑ *Left Hip*
❑ *Right Hand*	❑ *Left Hand*	❑ *Right Wrist*
❑ *Left Wrist*	❑ *Right Elbow*	❑ *Left Elbow*
❑ *Finger*	❑ *Abdomen*	❑ *Groin*

❑ *Right Foot*   ❑ *Left Foot*   ❑ *Toe*

❑ *Other (Specify)*   ❑ *Other (Specify)*   ❑ *Other (Specify)*

STEP 6 - DETERMINE ACCIDENT TYPE AND AGENCY

Accident Type (Contact with what Source of Energy)

Determine what exchange of energy caused the injury or damage. Once again, check a block and give a brief explanation.

Accident Type   **Explanation**

❑ *Struck against* _____

❑ *Struck by* _____

❑ *Caught in* _____

❑ *Caught on* _____

❑ *Caught between* _____

❑ *Slip* _____

❑ *Fall to same level* _____

❑ *Fall to lower level* _____

❑ *Overexertion* _____

❑ *Electricity* _____

❑ *Heat* _____

❑ *Cold* _____

❑ *Vibration* _____

❏ *Caustics/Acid* _____

❏ *Noise* _____

❏ *Toxic/non-toxic substance* _____

❏ *Foreign object* _____

❏ *Other* _____

❏ *Other* _____

Agencies

The damage was caused by some article that was closely related to the damage or the injury. This is called the *agency*. Identify which general agency caused the loss. Check a block and give a brief explanation.

General Agency   **Description**

❏ *Installations* _____

❏ *Equipment* _____

❏ *Machinery* _____

❏ *Sharp Edge* _____

❏ *Container* _____

❏ *Surface* _____

❏ *Projectile* _____

❏ *Compressed Gas Cylinders* _____

❏ *Building/Structure* _____

❏ *Ladder/Stairs* _____

❑ *Obstruction* _____

❑ *Explosive Device* _____

❑ *Animal/Insects* _____

❑ *Other* _____

❑ *Other* _____

List any occupational hygiene agencies that may have contributed to the loss.

Hygiene Agency   **Description**

❑ *Chemical* _____

❑ *Dust* _____

❑ *Fumes/Vapors* _____

❑ *Noise* _____

❑ *Fire* _____

❑ *Gas* _____

❑ *Heat* _____

❑ *Radiation* _____

❑ *Biological* _____

❑ *Ergonomics* _____

❑ *Lighting* _____

❑ *Other* _____

❑ *Other* _____

STEP 7 - DESCRIPTION OF THE EVENT

By now, enough information has been gathered for a brief but accurate description of the near-miss or accident. While writing this, do not attempt to pose solutions, give your personal views, or make any assumptions. Give only the facts of the accident. If people committed unsafe acts it is a fact, if there were unsafe conditions, it is a fact.

DESCRIBE CLEARLY WHAT HAPPENED:

_____

_____

_____

_____

_____

_____

_____

_____

_____

_____

_____

_____

_____

_____

_____

_____

_____

_____

_____

_____

STEP 8 - DETERMINE SEQUENCE OF EVENTS

Now that we have a description of the event it is time to put together a chronological sequence of events that occurred that led to the contact and subsequent losses. Refer to witnesses' interviews, and other reports for this information.

Number   **Time**   **Date**   **What happened?**

_____

_____

_____

_____

_____

_____

_____

_____

_____

_____

_____

_____

_____

_____

_____

_____

_____

STEP 9 - DO AN IMMEDIATE CAUSE ANALYSIS

It is now time to determine what unsafe conditions contributed to the contact. Unsafe conditions are the factors within the physical environment that were unsafe and that posed a hazard. Unsafe conditions are the immediate cause of the accident and can be categorized under certain headings. Check a block and give a description of the unsafe conditions that existed at the accident site.

Unsafe Condition   **Description**

❑ *Inadequate Guard or Barrier* _____

❑ *Inadequate or improper PPE* _____

❑ *Defective Tools/Equipment/Materials* _____

_____

❑ *Congested/Restricted/Overcrowded Work Area* _____

_____

❑ *Inadequate warning system* _____

❑ *Fire and explosion hazard* _____

❑ *Poor housekeeping* _____

❑ *Hazardous environment* _____

❑ *Noise exposure* _____

❑ *Radiation exposure* _____

❑ *High or low temperature exposure* _____

❑ *Inadequate/excessive illumination* _____

❑ *Inadequate ventilation* _____

❑ *Other (specify)* _____

❑ *Other (specify)* _____

To complete the immediate cause analysis the unsafe acts must now be identified. Check the appropriate unsafe act(s) that was evident as a result of the investigation and give a brief description where necessary.

Unsafe Act   **Description**

❑ *Operating equipment without authority/training* _____

_____

❑ *Failure to warn* _____

❑ *Failure to secure* _____

❑ *Operating at improper speed* _____

❑ *Making safety devices inoperable* _____

❑ *Removing safety devices* _____

❑ *Using defective equipment* _____

❑ *Using equipment improperly* _____

❑ *Failure to wear PPE* _____

❑ *Improper loading* _____

❑ *Improper placement* _____

❑ *Improper lifting* _____

❑ *Improper positioning* _____

❑ *Servicing equipment in operation* _____

❑ *Horseplay* _____

❑ *Not following procedure* _____

❏ *Other (specify)* _____

❏ *Other (specify)* _____

❏ *Other (specify)* _____

STEP 10 - DO A BASIC CAUSE ANALYSIS

The unsafe acts and unsafe conditions are the immediate causes of the accident but are only the symptoms. The real causes are the basic causes that contributed to the unsafe act and unsafe condition. These have often been called the root causes and their identification is vital to the investigation. These are the causes that must be addressed. Action plans must be drawn up to prevent them from recurring.

Basic cause analysis normally opens a can of worms. During this analysis, courage and tenacity are needed if the investigation is to be done correctly and solutions to the real problems are to be found.

The key to deriving basic accident causes is to examine each unsafe act and condition and ask the questions why? why? why? The resulting answers will be either the personal or job factor that led to the unsafe practice and/or condition.

PERSONAL FACTOR ANALYSIS

A personal factor analysis is a method of deriving the personal factors that contributed to the person committing an unsafe act. Refer to each unsafe act listed in this investigation and ask the basic cause analysis questions. This will identify personal factors.

Lack of Knowledge

Ask the following questions:

❑ *Did the person have adequate knowledge to do the task that he was doing?*

❑ *When last was the procedure for the task reviewed with the person?*

❑ *Could further training and instruction have prevented the person from committing the unsafe act?*

If there was any hesitation as to this person's knowledge then one of

the basic causes would be **a lack of knowledge.**

❑ *Why was it a lack of knowledge?* _____

_____

_____

Lack of Skill

Ask the following questions:

❑ *Was this person skilled enough to carry out the task?*

❑ *Was the person familiar with the equipment, machinery, and method for the task?*

❑ *If it is an unusual task, were proper instructions given before the task?*

If there were any doubts as to the person's skill, the basic cause could

be a **lack of skill.**

❑ *Why was it a lack of skill?* _____

_____

_____

Inadequate Capability

Ask the following questions:

❏ *Was this person physically capable of doing this task at the time of the accident?*

❏ *Was this person mentally capable of doing this task at the time of the accident?*

❏ *Is there any reason to believe that this person should not be doing this task?*

In reviewing the questions if there is any doubt then **inadequate capability** is a basic cause.

❏ *Why was there a lack of capability?* _____

_____

_____

Lack of Motivation (Improper Motivation)

Ask the following questions:

❏ *Was this person influenced by fellow workers or peer pressure?*

❏ *Why did the person deviate from the safe procedure?*

❏ *Was the person trying to save time by working in a hurry and taking short cuts?*

If these questions cause any doubt then a basic cause is a **lack of** or **improper motivation.**

❑ *Why was there a lack of or improper motivation?* _____

_____

_____

Stress

Ask the following questions:

Was this person under stress due to:

❑ *Peer pressure*

❑ *Pressure of work*

❑ *Family problems*

❑ *Financial problems*

❑ *Drug or alcohol dependency*

If any of the answers to the questions are positive, then there is **stress** involved as a basic cause.

❑ *Why was stress involved?* _____

_____

_____

Inadequate Leadership and Supervision

Ask the following questions:

❑ *Had the unsafe act been seen and condoned in the past?*

❑ *Was supervision aware of deviations from safe practices?*

❑ *Had supervision taken action to prevent unsafe behavior in the past?*

❑ *Was the injured person aware of work procedures?*

❑ *Could better supervision have prevented the accident?*

If any of the questions above leave doubt as to the adequate supervision, then there could be a **lack of leadership or supervision.**

❑ *Why was there a lack of supervision and leadership?* _____

_____

_____

_____

_____

JOB FACTOR ANALYSIS

A job factor analysis is where each unsafe condition that was found is subject to a series of questions and the basic cause for these unsafe conditions is identified. The same method is used for a *job factor* analysis as was used for the *personal factor* analysis.

Inadequate Tools and Equipment

Ask the following questions:

❑ *Were the correct tools being used?*

❑ *Were the tools in a safe condition?*

❑ *Were lockout procedures complied with?*

❑ *Are defective tools reported?*

If any of the answers are negative, this indicates that there were

inadequate tools and equipment.

❑ *Why were the tools and equipment inadequate?* _____

_____

_____

_____

_____

Inadequate Maintenance

Ask the following questions:

❑ *Is there a written maintenance schedule?*

❑ *Was maintenance of the equipment done?*

❑ *Was a maintenance checklist used?*

❑ *Were the correct parts used for the repairs?*

❑ *Would adequate maintenance have prevented the unsafe condition?*

In reviewing the answers, did they indicate improvement was needed? If so there was **inadequate maintenance.**

❑ *Why was there inadequate maintenance?* _____

_____

_____

_____

_____

Wear and Tear

Ask the following questions:

❏ *Were the tools used in the wrong way?*

❏ *Was the equipment used beyond normal service life?*

❏ *Was the equipment operated incorrectly?*

❏ *Were people trained to use the tools, equipment, or facilities?*

These questions could indicate that **wear and tear** was a job factor.

❏ *Why was there wear and tear?* _____

_____

_____

_____

Misuse and Abuse

Ask the following questions:

❏ *Were the tools, equipment wrongly used?*

❏ *Were they overloaded, operated at excessive speed, or otherwise*

   abused?

If the answer to any of these questions is "yes" it would indicate

misuse and abuse as a job factor.

❑ *Why was there misuse and abuse?* _____

_____

_____

_____

Inadequate Purchasing Controls

Ask the following questions:

❑ *Were the items used correctly purchased as per specifications?*

❑ *Were they the right items for the application?*

❑ *Were adequate instructions supplied with the equipment or articles?*

❑ *Could improved purchasing controls have prevented the accident?*

The answers to the questions could indicate if **purchasing controls**

are a basic cause.

❑ *Why was there a lack of purchasing controls?* _____

_____

_____

Inadequate Work Procedures or Practices

Ask the following questions:

❑ *Was a written procedure available?*

❑ *When last was the procedure updated?*

❑ *When last was the person instructed on the procedure?*

❑ *Is the procedure still applicable?*

❑ *Has the procedure ever been enforced in the past?*

❑ *Is the task a critical task and has it been analyzed and a written safe work procedure (WSWP) been written?*

If there was a weakness in the **procedures** or **practices**, this could be a basic cause of the accident.

❑ *Why were procedures or practices inadequate?* _____

_____

_____

Inadequate Engineering Control

Ask the following questions:

❑ *Were the hazards highlighted in the initial hazardous analysis?*

❑ *Why did inspections not identify the hazards?*

❑ *Was a hazard-identification exercise completed?*

❑ *Were there inherent hazards in the design?*

If the answers to these questions indicate accordingly, then **inadequate engineering control** may have been a basic cause.

❑ *Why was it lack of engineering control?* _____

_____

_____

_____

STEP 11 - REVISIT THE IMMEDIATE CAUSE/BASIC CAUSE ANALYSIS AND INSPECT RECORDS

Revisit

At this stage, it is important to revisit the unsafe acts and conditions listed and for each unsafe act, ask the questions, why? why? why? and see whether all the basic causes have been identified. This is a critical part of the investigation procedure.

Repeat this with the unsafe conditions and ask the questions why? why? why? to ensure that basic causes have been found.

Review the basic cause analysis to make sure that you have delved deeply into the reason for the immediate causes of the accident. At this stage, do not try to pose solutions but merely identify the deeply hidden problems that caused this event. Wherever possible, justify why you indicated that a particular basic cause contributed to an unsafe act or unsafe condition.

Records

In answering the basic cause analysis questions you may have to inspect reports such as induction training, critical task instruction, pre-employment selection practices and other documents and evidence. These may help identify basic causes. This may take a while, but once again, it must be emphasized that to do a thorough accident investigation, all avenues must be explored.

In most accident and incident investigations there are normally more basic causes than there are immediate causes. In some instances up to six basic causes have led to the creation of one unsafe condition. Finding these contributory factors is vital to this investigation process.

STEP 12 - RECOMMEND CONTROL REMEDIES

Now that the basic causes of the accident have been identified, the necessary control measures or actions must be put in place to re-establish the breakdown in the safety management system that caused them. Only if positive remedial measures are instituted can a recurrence of a similar accident be prevented.

Important

The proposed actions to eliminate the accident causes are **the most important** step in the accident investigation process. After reviewing each basic cause under the headings of personal factors and job factors, institute controls that will prevent them from creating an unsafe act or an unsafe condition in the future.

Some actions may be immediate, some may be medium-term and some may be long-term actions. An example of a long-term action is the introduction of a safety committee system. This will involve the compiling of the constitution, the organizational structure of the committees, obtaining the permission of management, arranging the meetings, etc.

Treat the Cause

Many accident investigators make the mistake of treating only the symptoms of the accident and they never eliminate the real causes. Ensure that the control steps address the real causes of the accident as shown by the basic cause analysis.

Control Steps

Each control step should have 4 parts:

1. What action must be taken?

2. Who is responsible for taking this action?

3. By what date must this action be completed.

4. A follow up to ensure the action has been successfully executed.

The control step action plan is as follows:

Control Steps to prevent recurrence	**By whom**	**Date to be completed**	**Date completed**
1.			
2.			
3.			
4.			
5.			
6.			
7.			
8.			
9.			
10.			

Work Request/Maintenance Order

Some of the action steps indicated may require specific work orders or action from maintenance departments. These can be initiated by the following table, which will also enable you to keep a record of the work orders.

Work Orders	**Sent to Whom**	**Action Date**	**Completion Date**
1.			
2.			

3. _____

4. _____

5. _____

STEP 13 – FOLLOW-UP

Follow-up actions to monitor the effectiveness of the corrective actions are just as important as the corrective actions themselves. This follow-up should be done before the accident investigation has been closed and completed. If there were mechanical or engineering remedies to the accident problem, a thorough inspection of these in place will measure whether they are effective and will prevent a recurrence.

If administrative controls in the form of training, retraining, or adjusting work tempos and positions are needed, these should also be monitored and tracked on a follow-up action control sheet.

The follow-up control sheet is as follows:

What Follow-up Actions were Implemented   **Date Completed**
to Monitor the Effectiveness of Corrective
Actions?

1. _____

2. _____

3. _____

4. _____

5. _____

6. _____

7. _____

8. _____

9. _____

10. _____

<u>STEP 14 - SIGNATURES</u>

Once the follow-up actions have been monitored and completed and you feel confident that they will close the gap and prevent a recurrence of a similar accident, the necessary signatures are required. The investigation report should now be routed through the various levels of management for their comment and signature.

Investigator

The investigator should be the first person to sign off and he should sign the report only if all the control measures have been implemented.

"I find this report satisfactory, completed, and, to the best of my ability, I have ensured that all control measures to prevent this type of accident happening again have been implemented."

Investigator's Name (print) _____

Signature _____ *Date* _____

Area Supervisor's Name (print) _____

Signature _____ *Date* _____

Area Safety and Health Representative's Name (print) _____

_____

Signature _____ *Date* _____

One-up Manager

My comments on this accident investigation are:

_____

_____

_____

_____

_____

Name (print) _____ *Signature* _____

Date _____ *Routed to* _____

One-up Manager

My comments on this accident investigation are:

_____

_____

_____

_____

_____

Name (print) _____ *Signature* _____

Date _____ *Routed to* _____

One-up Manager

My comments on this accident investigation are:

_____

_____

_____

_____

_____

_____

Name (print) _____ *Signature* _____

Date _____ *Routed to* _____

General Manager

My comments on this accident investigation are:

_____

_____

_____

_____

_____

_____

Name (print) _____ *Signature* _____

Date _____ *Routed to* _____

Safety Department

Comments

_____

_____

_____

_____

_____

_____

_____

_____

Name (print) _____ *Signature* _____

Date _____

CONCLUSION

By now you should have completed a thorough, effective, and worthwhile investigation of the accident. If all the steps have been followed as described, the implementation of the control steps will certainly mend the breakdown in the management system and reduce the potential for a similar accident occurring due to the same basic causes.

This accident investigation report, in certain instances, is a legal document and should be treated as such and retained in a secure place for the period prescribed by your local safety and health legislation.

You may question the fact that the safety department is the last department to comment on the investigation. An accident investigation is management's responsibility, as it was a failure of the management system that caused the accident and consequent loss. Safety coordinators can be used to assist in the investigation but the investigation should not be their responsibility unless they have been specially assigned to the investigation for a specific reason.

It is unfortunate that so much effort is put into investigating an event that has already happened and has caused pain and suffering and loss to the organization. Nevertheless, an accident is an opportunity to learn a lesson that will prevent similar pain, suffering, and loss in the future. It is worthwhile doing a good job of accident investigation to prevent the sequence of events recurring. The best method is to identify the risks, assess them and put in the controls before an accident occurs.

SOURCES

- *Questions and Answers on Effective Accident/Incident Investigation,* National Occupational Safety Association, ISBN No. 0-620-13905-6.

- *Contemporary Issues in Strategic Management,* Elsabie Smit et al., Chapter 11, The Strategic Implications of Safety, R.C. McKinnon, ISBN No. 0-7986-4612-8.

- *Practical Loss Control Leadership,* Frank E. Bird Jr. and George L. Germain, International Loss Control Institute, Inc., ISBN No. 0-88061-054-9.

- *A Guide to Effective Accident/Incident Investigation and Causal Analysis,* Western Deep Levels, South Mine (unpublished document).

- *The Cause, Effect, and Control of Accidental Loss (CECAL)©,* R.C. McKinnon.

INDEX

Printed and bound by CPI Group (UK) Ltd, Croydon, CR0 4YY

23/10/2024

01778249-0009